Global Security & Intelligence Studies

Also from Westphalia Press
westphaliapress.org

Global Security & Intelligence Studies

Volume 4, Number 2
Fall/Winter 2019

Melissa Layne, Editor

WESTPHALIA PRESS
An imprint of Policy Studies Organization

Global Security and Intelligence Studies: Volume 4, Number 2, Fall/Winter 2019
All Rights Reserved © 2019 by Policy Studies Organization

Westphalia Press
An imprint of Policy Studies Organization
1527 New Hampshire Ave., NW
Washington, D.C. 20036
info@ipsonet.org

ISBN-13: 978-1-63391-904-4

Cover design by Jeffrey Barnes:
jbarnesbook.design

Daniel Gutierrez-Sandoval, Executive Director
PSO and Westphalia Press

—————————————

Updated material and comments on this edition
can be found at the Westphalia Press website:
www.westphaliapress.org

Global Security and Intelligence Studies
Volume 4, Number 2, Fall/Winter 2019
© 2019 Policy Studies Organization

GLOBAL SECURITY
AND INTELLIGENCE STUDIES
JOURNAL

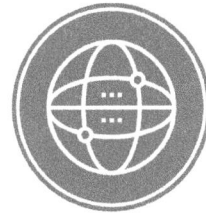

Editorial Welcome

Welcome to the Fall/Winter 2019 issue of GSIS. This second issue of the year features three original research articles, a unique perspective from a field setting, and research note, commentary on a policy-relevant issue, and two book reviews. Together, the research articles examine a breadth of issues related to intelligence and global security.

First, Keith Ludwick's "Lone Wolf Terrorism in Legislation: A Legal Overview" provides in in-depth look at the inconsistent laws addressing lone wolf terrorism that still exist today, as well as a discussion defining lone wolf terrorism. An interesting historical timeline examines terrorism laws pre- 9/11 and post 9/11. Given the current landscape regarding terrorism, this article is timely, and a must-read.

The second article, "The Neglected Dimension of Ideology in Russia's Political Warfare Against the West" by Armin Krishnan begins by reminding us of the ideological conflict between Russia and the West during The Cold War. This look back in history sets the stage for the introduction of Russia's new ideological social model, which serves as a viable vision that appeals to larger social groups. One of Russia's desired outcomes from this "war of ideas" is that the West seriously engage in this effort. This piece is an excellent example of countries working together despite differing political perspectives.

The third article, "Cyber Force Establishment: Defence Strategy for Protecting Malaysia's Critical National Information Infrastructure Against Cyber Threats" by research team Norazman Mohamad Nor, Azizi Miskon, Ahmad Mujahid. Ahmad Zaidi Zahri Yunos, and Mustaffa Ahmad, take proactive approach by proposing a conceptual framework toward addressing the critical need for cybersecurity threats in Malaysia. The framework suggests the establishment of a central agency to coordinate cybersecurity efforts between public and private sectors that manage the CNII sectors. Not only is this article a significant contribution to the literature, it is a significant contribution to other countries who may have similar issues.

Our fourth article, "Network Science in Intelligence: Intelligence Cell" by Romeo-Ionut Minican takes us into the fascinating world of secret and intelligence services. He shares how a basic hypothesis of graph theory may be used to identify the intelligence cell, information networks, the base circuit of information, the structure of secret services, the organization and the arrangement of safe houses, and the composition of network of informants - internal or external, on national, neutral or foreign territory.

In our fifth article, "Library of Spies: Building an Intelligence Reading List That Meets Your Needs", author Erik Kleinsmith provides the lifelong learner with a wealth of resources on where to find excellent reading lists on intelligence history,

 doi: 10.18278/gsis.4.2.1

intelligence, national security, espionage, and the more technical aspects of intelligence. Kleinsmith asserts that those who are avid readers in their respective field "bring a wealth of knowledge to their work…[and are] better able to understand and adapt themselves to different problem sets encountered on a regular basis." One book in particular he recommends, "The History of Spies and Spying" provides a look a covert operations, the origins of various intelligence organizations and agencies, espionage and traitors. Speaking of books, this makes for a perfect segue for our next two book review contributions.

Justin West provides an unbiased review of *Humanitarian Aid, Genocide, and Mass Killings: Medecine Sans Frontieres, the Rwandan Experience, 1982-97* written by authors Jean-Hervé Bradol and Marc Le Pape (2017). Without giving too much away, the authors provide the reader with a first-hand account of the trials and tribulations they experienced as MSFs working in Rwanda in the 80s and 90s. The collection of artifacts used to document these experiences are incredible.

And finally, Joel Wickwire takes a look at "International Organizations and the Law" written by Andrea Harrington (2018). Again, without disclosing his excellent review, I'll simply mention that the book examines the largely controversial history to present-day infrastructure of International Organizations (IOs). Harrington asserts that for IOs to reclaim "legitimacy" they must not only understand existing underlying laws, they must also work toward a flexible and innovative approach.

We are confident you will enjoy this exciting issue. I'd like to thank our talented authors and reviewers for sharing their valuable contributions with the Global Security and Intelligence Studies community.

Kind regards,

Melissa Layne, Ed.D.
Editor-in-Chief *Global Security and Intelligence Studies*

Le doy la bienvenida a la edición Otoño / Invierno 2019 de GSIS. Este segundo número del año presenta tres artículos de investigación originales, una única perspectiva desde un entorno de campo, y nota de investigación, comentario sobre un tema relevante para las políticas, y dos reseñas de libros. Juntos, los artículos de investigación examinan una variedad de temas relacionados con la inteligencia y la seguridad global.

Primero, el "Terrorismo del lobo solitario en la legislación: una descripción general legal" de Keith Ludwick proporciona una mirada en profundidad a las leyes inconsistentes que abordan el terrorismo del lobo solitario que todavía existen en la actualidad, así como una discusión que define el terrorismo del lobo solitario. Una interesante línea de tiempo histórica examina las leyes de terrorismo anteriores al 11 de septiembre y posteriores al 11 de septiembre. Dado el panorama actual con respecto al terrorismo, este artículo es oportuno y debe leerse.

El segundo artículo, "La dimensión desatendida de la ideología en la guerra política de Rusia contra Occidente", de Armin Krishnan, comienza recordándonos el conflicto ideológico entre Rusia y Occidente durante la Guerra Fría. Esta mirada hacia atrás en la historia prepara el escenario para la introducción del nuevo modelo social ideológico de Rusia, que sirve como una visión viable que atrae a grupos sociales más grandes. Uno de los resultados deseados de Rusia de esta "guerra de ideas" es que Occidente se compromete seriamente en este esfuerzo. Esta pieza es un excelente ejemplo de países que trabajan juntos a pesar de las diferentes perspectivas políticas.

El tercer artículo, "Establecimiento de la fuerza cibernética: estrategia de defensa para proteger la infraestructura de información nacional crítica de Malasia contra las amenazas cibernéticas", del equipo de investigación Norazman Mohamad Nor, Azizi Miskon, Ahmad Mujahid. Ahmad Zaidi

Zahri Yunos y Mustaffa Ahmad adoptan un enfoque proactivo al proponer un marco conceptual para abordar la necesidad crítica de amenazas de ciberseguridad en Malasia. El marco sugiere el establecimiento de una agencia central para coordinar los esfuerzos de seguridad cibernética entre los sectores público y privado que administran los sectores de la CNII. Este artículo no solo es una contribución significativa a la literatura, es una contribución significativa a otros países que pueden tener problemas similares.

Nuestro cuarto artículo, "Network Science in Intelligence: Intelligence Cell" de Romeo-Ionut Minican, nos lleva al fascinante mundo de los servicios secretos y de inteligencia. Comparte cómo se puede utilizar una hipótesis básica de la teoría de grafos para identificar la célula de inteligencia, las redes de información, el circuito base de información, la estructura de los servicios secretos, la organización y la disposición de las casas de seguridad, y la composición de la red de informantes. interno o externo, en territorio nacional, neutral o extranjero.

En nuestro quinto artículo, "Biblioteca de espías: construyendo una lista de lectura de inteligencia que satisface sus necesidades", el autor Erik Kleinsmith brinda al alumno de por vida una gran cantidad de recursos sobre dónde encontrar excelentes listas de lectura sobre historia de inteligencia, inteligencia, seguridad nacional, espionaje y los aspectos más técnicos de la inteligencia. Kleinsmith afirma que aquellos que son lectores ávidos en su campo respectivo "aportan una gran cantidad de conocimiento a su trabajo ... [y] están mejor capacitados para comprender y adaptarse a los diferentes conjuntos de problemas que se encuentran regularmente". Un libro en particular recomienda "La historia de los espías y el espionaje" ofrece una mirada a las operaciones encubiertas, los orígenes de varias organizaciones y agencias de inteligencia, espionaje y traidores. Hablando de libros, esto hace una transición perfecta para nuestras próximas dos contribuciones de reseñas de libros.

Justin West ofrece una revisión imparcial de ayuda humanitaria, genocidio y asesinatos en masa: Medecine Sans Frontieres, The Rwandan Experience, 1982-97, escrito por los autores Jean-Hervé Bradol y Marc Le Pape (2017). Sin revelar demasiado, los autores proporcionan al lector una descripción de primera mano de los ensayos y tribulaciones que experimentaron como MSF trabajando en Ruanda en los años 80 y 90. La colección de artefactos utilizados para documentar estas experiencias es increíble.

Y finalmente, Joel Wickwire echa un vistazo a "Las organizaciones internacionales y la ley", escrito por Andrea Harrington (2018). Nuevamente, sin revelar su excelente reseña, simplemente mencionaré que el libro examina la historia en gran parte controvertida de la infraestructura actual de las Organizaciones Internacionales (OI). Harrington afirma que para que las IO reclamen "legitimidad" no solo deben comprender las leyes subyacentes existentes, sino que también deben trabajar hacia un enfoque flexible e innovador.

Estamos seguros de que disfrutará de este emocionante tema. Me gustaría agradecer a nuestros talentosos autores y revisores por compartir sus valiosas contribuciones con la comunidad de Estudios de Seguridad e Inteligencia Global.

Saludos cordiales,

Melissa Layne, Ed.D.
Editora Principal *Global Security and Intelligence Studies*

欢迎阅读2019年《全球安全与情报研究》 (GSIS) 秋冬季期刊。作为今年第二期刊物，本期聚焦于三篇原创研究文章、一篇来自相关领域的独特视角文章、一篇研究报告、一篇与政策相关议题有关的评论文、以及两篇书评。总的来说，研究文章检验了一系列与情报和全球安全相关的问题。

第一篇文章《针对独狼恐怖主义的立法：一次立法回顾 》由作者Keith Ludwick撰写。文章深入研究了一系列不一致的法律，它们用于应对如今依旧存在的孤狼恐怖主义，文章同时探讨了对孤狼恐怖主义的定义。有趣的是，历史时间线分别检验了911袭击事件前后的恐怖主义法。考虑到当前的恐怖主义局面，这篇文章是一篇及时的必读文章。

第二篇文章《俄罗斯对西方的政治战中被忽视的意识形态维度》由作者Armin Krishnan撰写。文章开头提醒人们关注冷战时期俄罗斯与西方的意识形态冲突。回顾该历史有助于引入俄罗斯的新意识形态社会模式，这种模式作为一种可行的愿景，能吸引更大的社会团体。俄罗斯期望从"观念战"中实现的其中一个结果则是，西方严肃参与其中。就各国在持有不同的政治视角下共同努力而言，这篇文章堪称典范。

第三篇文章《网络力量的建立：保护马来西亚关键国家信息基础设施抵御网络威胁的防御战略》由Norazman Mohamad Nor, Azizi Miskon, Ahmad Mujahid Ahmad Zaidi, Zahri Yunos, Mustaffa Ahmad 五人组成的研究小组完成。这篇文章率先提出一项概念框架，用于应对马来西亚网络安全威胁的关键需求。框架提出建立一个中央机构，以协调那些管理CNII各部门的公共部门和私人部门就网络安全进行相关协作。

第四篇文章《网络情报科学：情报小组》由作者Romeo-Ionut Minican撰写。文章研究了情报机关的精彩世界。作者描述了图论这一基本假设如何可能被用于识别情报小组、信息网络、信息基本回路、情报机关结构、安全避难所的组织和布置、以及在国家领土、中立领土、或外国领土的内外部告密者网络的构成。

第五篇文章《间谍图书馆：建立一个满足你需求的情报阅读清单》中，作者Erik Kleinsmith 针对去哪找到有关情报历史、情报、国家安全、间谍活动、情报的技术方面的优秀阅读清单，为终生学习者提供了一笔资源财富。Kleinsmith 断言，那些在各自领域内热衷阅读的读者能"给工作带来知识财富，并且也更能理解不同问题，更能将自身适应定期遭遇的不同问题集。"Kleinsmith尤其推荐的一本书为《间谍和间谍活动的历史》 。这本书审视了秘密活动、不同情报组织和机构的起源、间谍活动、以及叛变者。提到书籍，接下来正好将介绍两篇书评。

作者Justin West客观地评论了由Jean-Hervé Bradol 和Marc Le Pape共同撰写的《人道主义援助、种族屠杀、大规模屠杀：1982年至1997年无国界医生在卢旺达的经历》 (2017) 。简短而言，这部书的作者为读者提供了上世纪八九十年代他们作为无国界医生在卢旺达工作时经历的审判和磨难的第一手记录。那些用于记录这段经历而收集的人工制品尤为震撼。

最后，作者Joel Wickwire评论了由Andrea Harrington撰写的《国际组织与法律》(2018) 。再次强调，在不透露书评的情况下，我将简短总结：这部书研究了从充满争议的历史到当下的国际组织 (IOs) 基础设施。Harrington断言，国际组织要想重获"合法性"，则必须理解现有基本法律，同时为灵活和创新的措施而努力。

我们相信您会享受阅读这期文章。我在此对将宝贵著作分享给GSIS社群的所有杰出作者和评审员表示感谢。

献上友好的问候，

Melissa Layne，教育学博士
《全球安全与情报研究》主编

Global Security and Intelligence Studies • Volume 4, Number 2 • Fall / Winter 2019

Lone Wolf Terrorism in Legislation: A Legal Overview

Keith W. Ludwick Ph.D.
American Military University, Charles Town, WV, USA

Abstract

Despite the frenzy of U.S. legislative activity addressing terrorism since the attacks of 9/11, laws specifically addressing lone wolf terrorism are still sporadic and inconsistent. This essay provides a legal overview of existing terrorism laws as they relate to political or religiously inspired violence conducted by an individual unaffiliated with an existing terrorist organization. It starts with a brief discussion on lone wolf terrorism followed by an examination of the history of terrorism laws. The body of the essay examines laws relating to lone wolf terrorism, dividing them into pre-9/11 and post-9/11. The article concludes with some recommendations as to why explicitly addressing lone wolf terrorism in U.S. legislation is essential for policymakers and academics.

Keywords: lone wolf, terrorism, laws, legislation

Terrorismo solitario en la legislación: una visión legal de conjunto

Keith W. Ludwick Ph.D.
American Military University, Charles Town, WV, EE. UU.

Resumen

A pesar del frenesí de la actividad legislativa de los Estados Unidos que aborda el terrorismo desde los ataques del 11 de septiembre, las leyes que abordan específicamente el terrorismo del lobo solitario todavía son esporádicas e inconsistentes. Este ensayo proporciona una visión general legal de las leyes de terrorismo existentes en relación con la violencia de inspiración política o religiosa realizada por un individuo no afiliado a una organización terrorista existente.

doi: 10.18278/gsis.4.2.2

Comienza con una breve discusión sobre el terrorismo del lobo solitario seguido de un examen de la historia de las leyes sobre terrorismo. El cuerpo del ensayo examina las leyes relacionadas con el terrorismo del lobo solitario, dividiéndolas en anteriores al 11 de septiembre y posteriores al 11 de septiembre. El artículo concluye con algunas recomendaciones sobre por qué abordar explícitamente el terrorismo del lobo solitario en la legislación estadounidense es esencial para los responsables políticos y académicos.

Palabras Clave: lobo solitario, terrorismo, leyes, legislación

针对独狼恐怖主义的立法：一次立法回顾

Keith W. Ludwick Ph.D.
American Military University, Charles Town, WV, USA

摘要

尽管美国自911袭击事件后进行了诸多立法活动，但针对独狼恐怖主义的相关立法依然很少且并不一致。本文从立法方面回顾了现有恐怖主义法，因为其有关于受政治或宗教引发的个体暴力，且个体不属于现有恐怖主义组织。本文首先简短探讨了独狼恐怖主义，之后检验了恐怖主义法的历史。正文部分研究了与独狼恐怖主义相关的法律，并将其划分为911事件之前和之后颁布的法律。为何在美国立法中明确研究独狼恐怖主义对决策者和学术界是必要的，本文结论就此提出了一些建议。

关键词：独狼，恐怖主义，法律，立法

Introduction

E ver since the USA PATRIOT Act brought counter-terrorism laws into the forefront of American minds, the debate regarding the efficacy of U.S. terrorism policy has grown. Fanning the flames, headlines publishing the latest congressional testimony of new/amended terrorism laws or questioning the "War on Terror" are common occurrences in the media (Turse 2018; Zimmerman 2018). Usually, the debate and discussion centers around what might be called "general" terrorism, especially groups who rely on religious ideologies. The public and policymakers typically create and apply laws, policies, and regulations to these types of violent groups. However, what about the specific issue of law and policy regarding lone wolf terrorism? Do legislative endeavors meet the needs of the United States Intelligence Community (USIC), law enforcement, and policymakers concerning this unique subset of terrorism? This paper aims to address that question by looking at existing U.S. laws and policies focusing on the narrow, and less often specifically debated, regulations and legal issues regarding lone wolf terrorism currently guiding the USIC and law enforcement community. What this study demonstrates is the legal apparatus at both the federal and state level is significantly lacking in addressing lone wolf terrorism.

Lone wolf terrorism continues to be the scourge of the early twenty-first century. Since 2000, the number of attacks conducted by individuals and unaffiliated groups has continued to grow (Worth 2016). As mentioned above, while numerous academic papers, articles, and other research venues devote significant pages and time to the study of lone wolf terrorism, there still lacks overarching policy addressing this specific threat. A noteworthy part of the reason for the absence of large-scale, broad U.S. policy toward lone wolf terrorism is due to two factors: lack of an accepted definition of lone wolf terrorism and the current governmental momentum for creating laws and policy addressing terrorist attacks conducted by large violent groups.

Over the past decade, various articles, essays, and books examined the definitional issue of lone wolf terrorism (Borum 2012; Ludwick 2016; Pantucci 2011). This still unsettled aspect of political violence presents some significant problems for policymakers, academics, and practitioners. Various researchers use the phrase lone wolf terrorism differently making research unstandardized and possibly confusing. If researchers are struggling to determine exact definitions, that will imply that legislators are equally struggling for consistency when creating new laws and policies. For purposes of this paper, lone wolf terrorism is an individual, or very small group, usually less than nine individuals, who commit or threaten the use of violence to influence a government for political or religious change, adapted from (Ludwick 2016).

This paper presents its analysis in several sections. It begins with a brief historical examination of terrorism in U.S. law for context. The next two sections divide the analysis of lone wolf terrorism laws into pre- and post-9/11. This division seems appropriate based on the significant shift in legislative activity after the attacks of 9/11 on the U.S. government and private sector. The public debate regarding terrorism laws arguably crescendoed with the passage of the USA PATRIOT Act forcing Congress and the Executive Branch to shift USIC legislation from focusing on intelligence collection to the development of legislation directly addressing terrorism. Finally, the paper will conclude with some recommendations and thoughts about the future of laws as they relate to lone wolf terrorism.

Impact of Analysis

A review of laws and policy is essential for several reasons. Many other disciplines have benefitted from annotated reviews of laws as they apply to particular topics. For instance, the discipline of cybersecurity helped from Fischer's report regarding the legal environment for cybersecurity laws in his paper by allowing researchers to compare new and existing laws against the nascent landscape of the cyber domain (Fischer 2014)"page":"77","source":"Zotero","event-place":"Washington D.C.","abstract":"For more than a decade, various experts have expressed increasing concerns about cybersecurity, in light of the growing frequency, impact, and sophistication of attacks on information systems in the United States and abroad. Consensus has also been building that the current legislative framework for cybersecurity might need to be revised.","number":"R42113","language":"en","author":[{"family":"Fischer","given":"Eric A"}],"issued":{"date-parts":[["2014",12,12]]}}}],"schema":"https://github.com/citation-style-language/schema/raw/master/csl-citation.json"} . Within terrorism studies, Zelman provides a solid overview of recent, general terrorism laws (Zelman 2001). From a different disciplinary perspective, Behuniak offers a review of humanitarian intervention by armed forces, and Wildenthal submits an overview of laws as they relate to Native American sovereignty (Behuniak 1978; Wildenthal 2017). This brief list is a small sampling of the hundreds and hundreds of articles demonstrating the usefulness of legal overviews. What is lacking is an overview of terrorism laws as they relate to lone wolf terrorists.

Second, providing a single resource for a reference regarding laws on specific topics benefits future researchers by allowing easier comparing and contrasting of relevant statutes. This paper fills this missing gap concerning U.S. legislation and policy concerning lone wolf terrorism. Scholars investigating this unique aspect of terrorism will benefit from the background information and the context demonstrating how the U.S. government legal corpus evolved to incorporate lone wolf terrorism.

History of Terrorism Laws

Current events frequently impact the development of legislation in the United States. One obvious notable example includes the civil rights movement during the 1960s precipitating new legislation guaranteeing equal protection of rights regardless of race. As such, it should not come as any surprise that the development of U.S. laws concerning intelligence collection and counter-terrorism followed the domestic intelligence activities of the 1970s and terrorist attacks of the late 1990s/early 2000s, respectively. This section illustrates how U.S. laws dealing with terrorism evolved out of the original laws which focused on counterintelligence. These new laws provided the USIC additional tools to investigate and stop terrorist activity. Unfortunately, during this shift in emphasis, the problem of dealing with a lone wolf terrorist was neglected. As such, what follows sheds light on existing law and policy to help evaluate whether they meet the needs of the USIC, law enforcement, and policymakers.

Federal Intelligence and Terrorism Laws—Pre-9/11

Terrorism laws before 9/11 were sparse and inconsistent at both the federal and state level. In 1935, the first mention of "terrorism" in the federal record occurred in a report entitled "To Punish for Exerting Mutinous Influence upon the Army and Navy":

> However, revolutions by violence and force, proceeding from impatience with the existing system of government, and originating with a small minority, and carried through by terrorism, wholesale murder, and universal confiscation, are essentially repugnant…("To Punish for Exerting Mutinous Influence Upon Army and Navy" 1935)

Although the meaning of terrorism in the above quote could be up for some interpretation, it represents a close enough connotation to the discussion as to be worth including as a demonstration of what is most likely the first reference to "terrorism" in U.S. law.

The first federal U.S. law specifically addressing "terrorism" in any form was the Foreign Assistance Act of 1969 (Foreign Assistance Act of 1969 1969). This law did not address terrorist's actions specifically but denied U.S. aid to the United Nations if organizations supporting terrorism received funding. Around the same time, another law, the International Security Assistance and Arms Export Control Act of 1976 regulating trade, prohibited interaction with countries which sponsored or condoned terrorism:

> SEC. 620A. PROHIBITION AGAINST FURNISHING ASSISTANCE TO COUNTRIES WHICH GRANT SANCTUARY TO INTERNATIONAL TERRORISTS.—(a) Except where the President

finds national security to require otherwise, the President shall ter-
minate all assistance under this Act to any government which aids
or abets, by granting sanctuary from prosecution to, any individual
or group which has committed an act of international terrorism…
(International Security Assistance and Arms Export Control Act of
1976 1976, vol. 2151, sec. 303).

What is interesting about this law is the deliberate inclusion of the phrase "…
individual or group…." The word "individual" provides a glimpse that lawmakers
of that time did not see terrorism as inherently a group activity. This view was not
necessarily incorporated into terrorism law and policy by present-day legislators.

From these first few references until 9/11, the inclusion of terrorism in federal laws increased slowly in the wake of several high-profile attacks including
those conducted by the Unabomber in the 1990s, the first World Trade Center
bombing in 1993, the Oklahoma City bombing in 1995, and the Kenya and Tanzania embassy bombings 1998. All the laws targeting terrorism during this time dealt
more with groups; individual terrorists were mentioned only occasionally and out
of the full context of the law. Interestingly, the first World Trade Center bombing
in 1993 did not inspire legislators to enact new laws addressing the threat. As noted
in the 9/11 Commissions report, the successful investigation, arrest, and prosecution of the individuals who perpetrated the first World Trade Center attack gave
the impression that the legal system was "…well-equipped to cope with terrorism"
(National Commission on Terrorist Attacks upon the United States 2004, 72).´

Foreign Intelligence Surveillance Act

When considering U.S. law and policy within the realm of intelligence and
terrorism, it is essential to include a brief explanation of the Foreign Intelligence Surveillance Act (FISA). This law is the bedrock for all intelligence activities conducted by the USIC on U.S. soil or against U.S. Citizens/perma-
nent residents. FISA history is steeped in the politics of the mid-to-late 1970s when
public debate arose regarding powers of U.S. agencies belonging to the Executive
Branch. When the Watergate and Church Commissions revealed questionable domestic and international intelligence activities by the Federal Bureau of Investigation
(FBI) and other federal agencies, the discussion regarding governmental legal processes (or more accurately, lack of legal process) for ongoing intelligence activities on
U.S. soil came to a head. In light of these commission's reports, one of the outcomes
was the passing of the Foreign Intelligence Surveillance Act (FISA) authorizing the
formation of the Foreign Intelligence Surveillance Court (FISC) (Foreign Intelli-
gence Surveillance Court | United States n.d.). This law laid out the legal framework
for how U.S. government agencies should obtain legal authority for all domestic, and
some foreign, intelligence activities. The critical component of this brief discussion
regarding the FISA centers on the issue of the focus on foreign intelligence activities,

not terrorist activities (and not even a mention of an individual terrorist). More specifically, it targets foreign powers or entities, defining them as:

A foreign power is—

1. a foreign government,

2. a diplomat, other representative or employee of a foreign government,

3. a faction of a foreign nation that is not substantially composed of U.S. persons,

4. an entity openly acknowledged by a foreign government to be directed and controlled by it, or

5. a *group* engaged in international terrorism or activities in preparation therefore (Foreign Intelligence Surveillance Act of 1978 1978).

A different section of the FISA discusses agents of foreign powers, and in paragraph (2) states:

An Agent of a foreign power is—

(2) any person who—

(A) knowingly engages in clandestine intelligence gathering activities for or on behalf of a foreign power, which activities involve or may involve a violation of the criminal statutes of the United States;

(B) pursuant to the direction of an intelligence service or network of a foreign power, knowingly engages in any other clandestine intelligence activities for or on behalf of such foreign power, which activities involve or are about to involve a violation of the criminal statutes of the United States;

(C) knowingly engages in sabotage or international terrorism, or activities that are in preparation therefore, for or on behalf of a foreign power; or

(D) knowingly aids or abets any person in the conduct of activities described in subparagraph (A), (B), or (C) or knowingly conspires with any person to engage in activities described in subparagraph (A), (B), or (C) (Foreign Intelligence Surveillance Act of 1978 1978, vol. 1801, sec. 101).

The law clearly emphasizes foreign nationals working on behalf of a foreign power and lacks detail about individual terrorists operating alone. Moreover, despite a definition of *international terrorism* in section 101 (c), there lacks discussion of an individual terrorist. This absent detail seems small but has tremendous implications; consider Zacharias Moussaoui's involvement (the assumed "20th hijacker") in the plot on September 11, 2001. As stated in a report from the Department of Justice Inspector General:

> The Minnesota FBI and FBI Headquarters differed as to whether a warrant could be obtained and what the evidence in the Moussaoui case suggested. FBI Headquarters did not believe sufficient grounds existed for a criminal warrant, and it also concluded that a FISA warrant could not be obtained because it believed Moussaoui could not be connected to a foreign power as required under FISA ("Special Report: A Review of the FBI's Handling of Intelligence Information Related to the September 11 Attacks (Full Report)" 2006, 101).

The lack of direct connection to a foreign power—Moussaoui did not "qualify" as an agent of a foreign power under the FISA definition—limited the USIC's, and more specifically, the FBI's, legal options. Closing this loophole might not have prevented the attack on the World Trade Center, but it illustrates a specific problem. These types of issues proved even more problematic as the United States entered the twenty-first century, and the focus of national security moved to the terrorist threat.

Looking broadly, back in the late twentieth century, legislators did provide law enforcement and the USIC some legal mechanisms to deal with terrorism, but often it seemed the legislation and policy struggled to "catch up" and adapt in the wake of increased terrorist attacks. From a legal perspective, existing laws and policy covered small groups because the law applies just the same as to a larger group if they can demonstrate linkage to an international organization, but lone wolf terrorism, or the distinction of individual terrorists versus those affiliated with groups, did not really appear within the terrorism lexicon before 1995. As such, laws and policy addressing this unique aspect of political violence did not exist. In contrast, individuals, especially those not working on behalf of a foreign power, would not fall under the purview of these old laws, so law enforcement based their prosecutions on criminal statutes treating the suspects no differently from "regular" criminals.

From an academic and research perspective, there are further problems. If scholars wish to seek out instances of terrorism when searching through legal records, the difficulty lies in determining prosecutions based on violations of terrorism statutes and those based on criminal laws. Since the primary goal in terrorism investigations for law enforcement is to mitigate the threat by any means, prosecuting individuals for a lesser charge, say money laundering or selling counterfeit goods, provides a more straightforward solution to remove a terrorist from the general population. Rather than waiting for enough evidence to prosecute someone under a terrorism statute and risk the individual conducting his/her attack, it is easier and quicker to charge and convict them of a more straightforward crime. This absence of discussion of this issue in the literature exemplifies the incomplete reporting, which fails to include all instances of terrorism, making studying and understanding difficult due to lack of data.

Analysis—Pre-9/11

What is essential about the above brief review of the history of terrorism laws is the lack of consideration of the threat from individuals operating alone who conspire or commit political and religious violence. Legislators did give some forethought to providing law enforcement and the USIC some legal mechanisms to deal with terrorism in general but failed to incorporate individual terrorists or lone wolves. From a legal perspective, laws consider small groups part of larger groups, but individuals, especially those who were not working on behalf of a foreign power, would not be included or even envisioned under the existing statutes. As a result, law enforcement and the Department of Justice's enforcement arms were forced to use laws already on the books addressing murder, arson, and bombings. These legal tools worked but did not specifically address the threat at hand. Consider this within the context of law enforcement and prosecutors tending to work from a perspective of "What can we charge an individual with?" and proceed with investigative steps to fulfill those prosecutorial goals. Law enforcement is lacking specific statutes targeting lone wolf terrorism and must search for other avenues from which to pursue these individuals and their threats.

Federal Intelligence and Terrorism Law—Post-9/11

USA PATRIOT Act

The defining period in legal policy regarding terrorism immediately followed the 2nd attack on the World Trade Center and the ensuing debate over the USA PATRIOT Act. The USA PATRIOT Act (formally: The Uniting and Strengthening America by Providing Appropriate Tools Required to Intercept and Obstruct Terrorism Act of 2001) laid the necessary groundwork and provided many useful tools for law enforcement and the USIC. Unfortunately, the USA PATRIOT Act failed to consider lone wolf terrorism and left gaps requiring later attention.

The USA PATRIOT Act was controversial almost from its inception. This public debate, a contentious discussion lasting about a month and a half, revolved around the limits of government power; it ultimately became law but incorporated a "sunset" clause of December 31, 2005, for several of its provisions (*P.A.T.R.I.O.T.*

Act of 2001 2001, vol. 1, sec. 224). While many pundits, legal experts, members of the media, and the general public argued for or against the passage of the act, the USIC was generally in favor of its enactment. There has been a significant discussion in the academic literature about the privacy implications of many aspects of the USA PATRIOT Act. For example, Doyle provides a neutral summary of the PATRIOT Act and examines the issue of privacy in the context of previous federal laws and how the interpretation of privacy has evolved over the years incorporating sanction against the government for violations of the PATRIOT Act (Doyle 2002). In Matz's article, the author examines specific issues regarding privacy, libraries, and educational institutions and suggests that the PATRIOT Act constitutes an overreach by the U.S. government (Matz 2008). Even media outlets such as CNN have provided insight into the privacy debate of the PATRIOT Act (CNN n.d.).

The USA PATRIOT Act aimed to provide law enforcement and the USIC better legal tools to counter and investigate terrorist groups targeting the United States. This law, which provided enhanced electronic surveillance procedures, money laundering abatement, and border protection, among other things, became law on October 26, 2001 (*P.A.T.R.I.O.T. Act of 2001* 2001). The debate about the USA PATRIOT Act would resurface in July 2005 as the sunset clauses were closing in on their expiration dates. In March 2006, Congress renewed the USA PATRIOT Act with few changes, none of which had to do with enhancing the U.S. government's ability to target lone wolf terrorism (*USA Patriot Improvement and Reauthorization Act of 2005* 2005). Although the USA PATRIOT Act did not address lone wolf terrorism, it laid the groundwork for the "Lone Wolf Amendment" to the FISA law three years later.

"Lone Wolf" Amendment

In July 2004, the National Commission on Terrorist Attacks Upon the United States ("The 9/11 Commission") publicly released its final report (National Commission on Terrorist Attacks upon the United States 2004, 11). The 9/11 Commission reviewed the actions of the USIC and FBI leading up to the attack on September 11, 2001, and developed recommendations to prevent future attacks. During this same period as the 9/11 Commission, attention focused on solving the issues brought up during Moussaoui's investigation by the FBI (as discussed earlier). All these topics coalesced to form the basis for the Intelligence Reform and Terrorism Prevention Act (IRTPA) of 2004. Often referred to as the "Lone Wolf Amendment," Section 6002 of the IRTPA focuses on lone wolf terrorism. Some of the changes the law implemented included the creation of the Director of National Intelligence and facilitating more sharing of FISA intelligence within the USIC and with law enforcement. The IRTPA also amended the FISA to fix the loophole regarding an individual unaffiliated with a foreign power who commits acts of international terrorism (Bazen and Yeh 2006, sec. 6001(C)). Before this change, the clause regarding an "agent of a foreign power" did not apply to individuals who acted alone and were not a part of a state-sponsored terrorist group inhibiting the law's application to terrorist investigations such as in the case of Moussaoui.

This IRTPA was not without its critics. Some felt the FISA, in its original interpretation, did allow the FBI to address unaffiliated individuals and considered it an overreach of law enforcement to target individuals without ties to foreign governments (DeRosa 2005). On December 31, 2009, the President signed the IRTPA into law (Bazen and Yeh 2006, 1). The FISA was amended to say:

SEC. 6001. INDIVIDUAL TERRORISTS AS AGENTS OF FOREIGN POWERS.

(a) IN GENERAL.—Section 101(b)(1) of the Foreign Intelligence Surveillance Act of 1978 (50 U.S.C. 1801(b)(1)) is amended by adding at the end the following new subparagraph:

"(C) engages in international terrorism or activities in preparation therefore; or".

(b) SUNSET.—The amendment made by subsection (a) shall be subject to the sunset provision in section 224 of Public Law 107–56 (115 Stat. 295), including the exception provided in subsection (b) of such section 224 (Bazen and Yeh 2006, sec. 6001).

As the original FISA law focused primarily on foreign intelligence activities, not terrorism, the IRTPA appeared to be a much-needed update to an outdated law. The IRTPA also signified a shift by Congress from creating laws dealing with intelligence collection by the USIC and beginning to include terrorism into future legislation systematically. The Assistant Attorney General at that time, David Kris, testified during a U.S. Senate hearing that the FBI had not used this provision in the five years since the passage of the law (*Statement of David Kris, Assistant Attorney General Before the Committee on the Judiciary, United States Senate Entitled "Reauthorizing the USA Patriot Act: Ensuring Liberty and Security"* 2009, 4). Kris argued in his testimony that the law offered a tool having application should a terrorist self-radicalize or "…severs his connection with his group…" (*Statement of David Kris, Assistant Attorney General Before the Committee on the Judiciary, United States Senate Entitled "Reauthorizing the USA Patriot Act: Ensuring Liberty and Security"* 2009, 5). Despite closing a loophole, there was an unforeseen problem concerning domestic terrorist acts. The IRTPA law states an individual who "…engages in *international* terrorism…" shall be included within the FISA law. The law fails to address those who commit domestic terrorism or where their relationship with a terrorist group, international or domestic, is not known.

The issue regarding domestic terrorism and U.S. laws are not usually a significant problem as most domestic terrorism investigations rarely use the FISA due mostly to the prohibition against targeting U.S. citizens for First Amendment issues protecting free speech. Traditionally, domestic terrorism investigations rely on criminal warrants. However, there have been instances of espionage or counterintelligence investigation "turning into" a domestic terrorism case or an international terrorist "becoming" a domestic terrorist because they obtain citizenship leading to calls for a "domestic terrorism" law (Tucker and Balsamo 2018). Since it is possible in only limited circumstances to use FISA law in domestic terrorism cases, the possibility exists that there would be difficulty in pursuing some sophisticated investigative techniques should they become necessary as it removes the "FISA tool from the toolbox" for investigators. Publicly available information has yet to reveal any examples of jeopardized investigations or prosecutions because of this loophole, but as is often the case, the problem will become evident due to some significant policy or prosecutorial failure and later become an issue.

Local/State Laws

As of this writing, there are no state laws and only one Resolution (brought to the floor in the Tennessee Congress in 2015) explicitly mentioning "lone wolf terrorism" (McNally 2015). Tennessee's Resolution admonishes the President of the United States and requests that he reconsider the attack by Nadal Hasan at Ft. Hood, Texas as a terrorist attack. Although searches for "terrorism" within state statues returned considerable results (1,000+), additional searches for "terrorist," "terrorist group," "terrorism AND individual," and "individual terrorist" yielded results unrelated to this discussion and addressed issues such as denying work permits, applications, or licenses to individuals associated with terrorism and did not deal specifically with the criminality of being a singular terrorist. A search of the municipal codes of New York City, NY, and Oklahoma City, OK—both cities exposed to significant terrorist events from individual terrorists—returned no results (American Legal Publishing—Online Library 2016; Municode Library 2015).

United States Intelligence Community and Department of Defense Policy

Thinking of federal laws as the topmost layer of regulation, policy by specific agencies within the U.S. government represents more detailed guidance to the respective employees of the various agencies on addressing terrorism. This section attempts to look at how individual agencies within the Executive Branch deal with lone wolf terrorism.

Federal Bureau of Investigation

Although the FBI's top priority is to stop terrorist attacks and protect the American public, the organization does not approach lone wolf terrorism as a subset of general terrorism. To the FBI, investigations into potential terrorist activities start with a complaint from the public or other local/state/federal agencies regarding an individual who is possibly engaged in politically or religiously motivated violence. From that point, investigators follow leads to develop a picture of how that individual is operating and if others might be involved with the potential threat. Each individual potentially involved in the activities is a stand-alone case arising out of an existing legal system where each terrorist would most likely be prosecuted independently of the other, so each case usually starts independently. To be clear, investigators do not work in a vacuum without coordinating efforts with others and their related cases. They also incorporate other types of intelligence or work together with cooperating witnesses. However, for the FBI, the thought of a lone wolf terrorist has little impact on investigative strategy or prosecutorial outcome.

The only element close to a lone wolf terrorist in the semantics of the FBI is the Homegrown Violent Extremist (HVE). The FBI has published many internal intelligence reports and articles about the threat of HVE, their origins, and mitigation strategies; some of these merit some consideration against lone wolf terrorism, but this approach looks at HVE in a different light, more from what types of origins HVE's come from rather than mitigation strategies (Homegrown Violent Extremism n.d.). Recently, in the winter of 2016, the FBI began a media campaign to make the public more aware of the warning signs of HVE's, but this has little to do with specifics of lone wolf terrorism (FBI | Countering Violent Extremism n.d.).

Department of Homeland Security

The Department of Homeland Security (DHS), as the newest member of the USIC, has its roots in preventing terrorism, having formed in the aftermath of the attacks of 9/11. However, because of its behemoth size, numerous agencies (as of this writing, 22 different agencies), and complex Congressional reporting requirements, locating specific policies regarding terrorism in general, let alone lone wolf terrorism, is difficult ("About DHS" 2012)000 employees in jobs that range from aviation and border security to emergency response, from cybersecurity analyst to chemical facility inspector. Our duties are wide-ranging, and our goal is clear - keeping America safe.","URL":"https://www.dhs.gov/about-dhs","language":"en","issued":{"date-parts":[["2012",5,4]]},"accessed":{"date-parts":[["2019",3,4]]}}}],"schema":"https://github.com/citation-style-language/schema/raw/master/csl-citation.json"} . The various law enforcement agencies within its structure, including Immigration and Customs Enforcement, the Coast Guard, and Secret Service, all have a specialized niche and address terrorism tangentially within the purview of their regular duties. Being federal law enforcement, they have all the same prosecutorial tools as the FBI as well as the ability to investigate terrorist crimes, but since the FBI is the lead federal agency charged with investigating terrorism, they usually do not incorporate terrorist investigations into their core investigative efforts.

Central Intelligence Agency

With the primary objective of collecting intelligence overseas, and legally prohibited from operating domestically, the Constitution and other laws rarely constrain the Central Intelligence Agency (CIA) while operating internationally. However, as an agency with the primary focus of collecting intelligence to inform policymakers, the actions of individuals, or even small groups, who conduct political or religious violence are of secondary concern. Of course, the CIA would have concerns regarding lone wolf terrorism and would not ignore an imminent attack by an individual working without being affiliated with a group, but the CIA operational model does not facilitate concentrating on this type of terrorism. The CIA approach focuses more on larger, more established terrorist organizations. Their definition of terrorism demonstrates this:

- The term "terrorism" means premeditated, politically motivated violence perpetrated against noncombatant targets by subnational groups or clandestine agents.
- The term "international terrorism" means terrorism involving the territory or the citizens of more than one country.
- The term "terrorist group" means any group that practices, or has significant subgroups that practice, international terrorism ("Central Intelligence Agency—Terrorism FAQ" n.d.).

While the CIA's definition of terrorism does include "clandestine agents," their efforts are more strategic, concentrating on groups and subgroups.

Department of Defense

Much like the CIA, the Department of Defense (DoD) does not have jurisdiction or operating authority within the United States except under limited circumstances. As such, similar to the CIA, its counter-terrorism approach is much more focused on larger, more organized threats. Although there are elements of DoD policy and guidelines which emphasize protecting the individual warfighter from an attack, these are not policies or guidelines directed at the more significant issue of lone wolf terrorism in general, but specific tactics to prevent or reduce injuries should an attack occur, regardless of the perpetrator. Domestically, from a security or "force protection" standpoint, DoD policies are focused on securing military bases and the individuals who live and work within them, best illustrated by the after-effects of the attack on the recruitment center in Chattanooga, TN, where the military vowed to increase security at recruiting stations (Miklaszewski, McClam, and Kube 2015).

Department of State

By statute, the Department of State (DoS) maintains the "official" list of terrorist groups recognized by the U.S. Government. Section 219 of the Immigration and Nationality Act directed the DoS to create and maintain the Foreign Terrorist Organization (Foreign Terrorist Organizations n.d.). This list, updated continuously, is used by the federal government as the official list of recognized terrorist organizations.[1] Although the list deals with both state-sponsored and stand-alone terrorist organizations, it does little to incorporate the uniqueness of smaller groups or individuals. Indeed, even the title, Foreign Terrorist Organizations, demonstrates the focus of the list.

The Foreign Terrorist Organization list is a convenient tool for governments and organizations to use as a benchmark when working to consider an organization affiliated with terrorism. Not surprisingly, the emphasis for the DoS is foreign organizations and does not incorporate domestic terrorist groups. One might also question its breadth due to the size of the list; as of this writing, the DoS lists 59 groups as Foreign Terrorist Organizations. In contrast, the Global Terrorism Database catalogs well over 1,000 unique groups (some of which are now defunct) (Global Terrorism Database 2018). Within the DoS website, lone wolf terrorism is not mentioned at all and is not incorporated into their guidance or policy.

Analysis—Post 9/11

The previous section aimed to describe current U.S. laws and policy as they relate to individuals who commit political or religious violence, the so-called lone wolf terrorists, after 9/11. Except one amendment to cover a loophole in federal law, the IRTPA, the discussion up to this point demonstrates that the legislators and policymakers failed to incorporate lone wolf terrorism, into the body of laws and guidelines used to catch and prosecute them. At some point, at least within the U.S. government, lawmakers realized the need for changes and implemented corrections and amendments to the existing FISA law, but this approach barely fixes a narrow scope of the problem, mostly from the intelligence side, and does nothing regarding criminal approaches to the problem.

Of course, the United States does have legal tools to prosecute those who commit these violent acts. To be sure, laws concerning weapons of mass destruction, trafficking in firearms, fraud, destruction of government property, and many others can be and are frequently used to prosecute terrorists and incarcerate them for significant periods. For example, the government charged Olympic Park Bomber Eric Rudolph with five counts of "...malicious use of an explosive..." and the government charged Faisal Shahzad (who attempted to explode a truck in New

York's Times Square) with five counts relating to the attempted use of a weapon of mass destruction ("#477: 10-14-98 Eric Rudolph Charged in Centennial Olympic Park Bombing", 1998; "Complaint Against Faisal Shahzad", 2010). Both received long sentences. However, laws used to convict these "lone wolf terrorists" were not the most optimum prosecutorial tool. They failed to address the underlying issue of a lack of understanding of the differences between lone wolf terrorism and terrorism committed by large groups and failed to identify that an attack committed by a single person is just as much an act of terror as if a group committed it. As mentioned above, if the legal system employs other laws to prosecute individual terrorists, it is difficult for policymakers and researchers to determine the full extent of lone wolf terrorism as an essential issue, mainly before the year 2001.

Recommendations

Overview

Suggestions for legal reform traditionally generate skepticism, doubt, and frustration. It is no secret that the U.S. system of government can most certainly bog down legal reform to the point of discussions and efforts becoming stagnant. However, some actions outside of traditional legal reforms will help fill the gaps with some of the legal issues discussed above. These recommendations fall into four categories: working to define lone wolf terrorism, reconsidering federal dominance, reduce legal ambiguity, and moving toward Executive Branch consistency. However, it is more than just "adding a few words" to a law. It requires analysis, thought, and dedication.

Define Lone Wolf Terrorism

While attempting to define terrorism in general is still under some debate, the consensus is building toward a standard definition centering around violence, or the threat of violence, by nonmilitary groups to influence governmental policy. However, as mentioned above, academics and policymakers still struggle to concur on an adequate definition of lone wolf terrorism. As many researchers have determined, even the name "lone wolf terrorist" is useless and misleading (Gruenewald, Chermak, and Freilich 2013; Hoffman 2016; Ludwick 2016). Other researchers try to avoid the issue altogether by using other labels such as "lone offender," "loner," "lone wolf pack," and "solo terrorist" (Borum 2012; LaFree 2013; Pantucci 2011). Unfortunately, these new labels only add further confusion to the definitional issue.

The lack of even a remotely agreed upon definition of lone wolf terrorism plays a central role in the development of laws and policy toward this terrorist threat. Creating standard definitions addressing terrorism by unaffiliated individuals will help scholars and policymakers in creating effective legislation. Scholars should continue to research and debate the definitional issue of unaffiliated politically violent individuals to help inform policymakers about the lack of an accepted definition, and, until there is a standard agreement, encourage language in laws that refrains from phrases like "lone wolf," "lone offender," and "solo terrorist." Creating standard definitions is not purely an exercise but will provide insight into which individuals are motivated by political or religious ideology instead of "lumping them in" with criminal activity. It is probably safe to assume that most researchers, legislators, and even the general public would consider acts of terrorism—especially those that result in a multitude of deaths and attempt to influence governments outside of the normal democratic process—much more severe of a crime than simple criminal activity such as bank robbery or murder. Creating laws to properly punish individuals who conduct these terrorist activities, rather than use simple criminal statues, requires an adequate definition.

As research grows, this will have the added benefit of others having more data to draw from; future research will undoubtedly benefit. Currently, it is difficult to understand the extent of lone wolf terrorism activity fully. Without a standard definition or the legislation to back it up, policymakers and researchers are having a difficult time even counting the number of lone wolf terrorists or the number of attacks they have committed.

Reconsider Federal Dominance

States and local branches of government need to reconsider federal dominance with respect to lone wolf terrorism. Often, state laws are used as "backups" to prosecute terrorists in the event issues arise with federal prosecution, or to "add time" to a terrorist's federal sentencing to ensure it is years before the perpetrator is set free. However, usually, these state laws tend to be criminal statutes such as money laundering, weapons possession, or murder. Again, as mentioned above, often the goal is simply to get the suspected terrorist off the street; in these instances, any criminal statute that applies will suffice.

However, state and local governments should consider how the various roles of prosecutorial bodies at the local, state, and federal level behave. The Department of Justice (DoJ) excels at prosecuting large scale, criminal conspiracy types of crimes. While they also prosecute individuals for lesser crimes, the experience of DoJ and their access to resources make them well suited for prosecuting complex crimes and terrorists. However, individuals who conduct terrorist attacks alone are much more likely to come to the attention of local and state law enforcement agencies before becoming known to federal agencies. Lone wolf terrorists rarely communicate with others with similar violent ideologies and often conceive, plan, and conduct their attacks within local jurisdictions without extensive travel beyond their local environment. It then makes sense to consider local and state law enforcement agencies as more likely to discover, investigate, and prosecute these individuals. As mentioned previously, terrorist activity is more likely to be dangerous to a local populace and result in death, so higher penalties should be warranted. States should develop and pass legislation which addresses these individual terrorists, so they move beyond merely charging and prosecuting individual terrorists with lesser crimes such as weapons possession or attempting to use explosive devices. States must move beyond relying on federal prosecution as the number of terrorist attacks conducted by unaffiliated individuals continues to grow.

Reduce Current Legal Ambiguity

As clearly demonstrated by the above discussion, current laws and policy have neglected the growing threat of lone wolf terrorism. It seems that legislatures at the federal and state level have failed to address the reality that unaffiliated individuals are conducting an increasing number of terrorist attacks. Laws need to be updated to ensure that the legal system moves beyond charging perpetrators with simplistic crimes such as bombings or murder when terrorism seeks to impact as wide as an audience as possible.

One possible solution would be to consider more substantial penalties for individuals who plan or conduct terrorist attacks alone. While this might seem counter-intuitive at first, there is a strong possibility that this approach could lead to more prosecutions of those terrorists affiliated with groups. If the deterrent effect of criminal penalties can be assumed to be effective, it is possible that some with a propensity toward terrorism might pursue groups when conducting terrorist activities to ensure a lesser penalty if ultimately caught. From an investigative standpoint, it has been well established that lone wolf terrorists are difficult to investigate for the simple fact that they are working alone (Phillips 2011). If stronger penalties could drive these types of individuals to work in terrorist organizations, then it is more likely that these individuals would be found out by law enforcement agencies or penetrated by informants leading to higher incidents of disruption and arrest. More research would undoubtedly be required to evaluate this approach, but this type of unique thinking is required when considering laws addressing lone wolf terrorists.

Move Toward Consistency

Executive Branch agencies inconsistently apply the existing laws addressing lone wolf terrorism, in large part due to the inconsistency in the laws themselves. Agencies like the FBI and DHS, neither which define "lone wolf terrorism," cannot create overarching policies helping them work together toward addressing lone wolf terrorism. While some agencies like the FBI do develop terms such as HVEs, which closely parallels the definition of lone wolf terrorism, the fact that the FBI creates their term only further illustrates the issue. When the "boots on the ground" investigators try to work together, they often find that their policies with addressing lone wolf terrorism do not mesh, which ultimately causes confusion and possible investigative problems. Further exacerbating the situation, law enforcement must enforce law created by Congress and prosecuted by the DoJ, both organizations which also have different ideas of what lone wolf terrorism is.

These issues can be alleviated quickly by creating multi-agency groups who can meet to develop definitional standards which can be forwarded on to Congress and the larger DoJ to help guide them in their development of policies and laws. If society acknowledges that investigators from these organizations have the expertise and experience to help develop definitions, with input from academics and scholarly sources, then standardized definitions to be used in the development of legislation should not be that difficult to realize. These efforts do not imply separate meetings and conferences should be convened solely for this purpose but incorporating these issues into the regularly scheduled counter-terrorism conferences and multi-agency meetings already held regularly can start the process.

Conclusion

Currently, law enforcement at all levels is stretched thin. While conducting their daily patrol and enforcement duties, many officers and investigators are more and more on the "front lines" of Homeland Security. Communities expect officers to help federal agencies such as the FBI by reporting suspicions of terrorist behavior and working with federal agencies in terrorism investigations. Couple this with the current trend of prosecuting terrorists for lesser charges and the public can expect many of these individuals to return to society posing a threat ultimately doubling the potential burden on those very same officers who are already spread so thin.

This paper provides background on how policy and laws inadequately incorporated elements of lone wolf terrorism, creating problems for legislators, policymakers, and practitioners. The unfortunate conclusion, under current policy and methodologies, is that lone wolf terrorism will always be a reactive issue, not a proactive one. The present understanding of lone wolf terrorism cannot support the sense of urgency placed on practitioners by the public and policymakers to stop all threats. Future researchers need to be mindful of the limitations of past and current laws when conducting their investigations and analysis. Their research should focus on three main areas: Continuing to refine the definition of lone wolf terrorism, or working to eliminate this outdated description altogether, evaluating how current laws have been implemented with respect to lone wolf terrorism to determine their efficacy, and attempting to put lone wolf terrorism in context of other criminal and terrorism statutes.

This paper demonstrates the current ambiguity in laws directed toward lone wolf terrorism hinders current and future policy. Scholars are encouraged to apply the concepts proposed in this paper to previous research efforts in order to determine if relevant findings need adjustment or refinement. Policymakers and law enforcement officials will benefit by considering the concepts presented and discussing these issues with legal professionals to determine the best path forward. While probably not eliminated in its entirety, part of reducing the threat of lone wolf terrorism to the United States lies in the hands of legislators.

References

"#477: 10-14-98 Eric Rudolph Charged in Centennial Olympic Park Bombing." 1998. Department of Justice. October 14, 1998. https://www.justice.gov/archive/opa/pr/1998/October/477crm.htm.

"About DHS." 2012. Department of Homeland Security. May 4, 2012. https://www.dhs.gov/about-dhs.

American Legal Publishing—Online Library. 2016. New York City Administrative Code. March 28, 2016. http://library.amlegal.com/nxt/gateway.dll/New%20York/admin/newyorkcityadministrativecode?f=templates$fn=default.htm$3.0$vid=amlegal:newyork_ny.

Barnes, Beau D. 2012. "Confronting the One-Man Wolf Pack: Adapting Law Enforcement and Prosecution Response to the Threat of Lone Wolf Terrorism." *Boston University Law Review* 92: 1613–33.

Bazen, Elizabeth B., and Brian T. Yeh. 2006. "Intelligence Reform and Terrorism Prevention Act of 2004: 'Lone Wolf' Amendment to the Foreign Intelligence Surveillance Act." RS22011. Congressional Research Service.

Behuniak, Thomas E. Captain. 1978. "The Law of Unilateral Humanitarian Intervention by Armed Force: A Legal Survey." *Military Law Review* 79: 157.

Borum, Randy. 2012. "A Dimensional Approach to Analyzing Lone Offender Terrorism." *Aggression and Violent Behavior* 17, no 5: 389–96. https://doi.org/10.1016/j.avb.2012.04.003.

Central Intelligence Agency—Terrorism FAQ. n.d. Accessed March 4, 2019. https://www.cia.gov/news-information/cia-the-war-on-terrorism/terrorism-faqs.html?tab=list-3.

CNN, Jeremy Diamond. n.d. "Patriot Act Debate: Everything You Need to Know." CNN. Accessed March 29, 2016. http://www.cnn.com/2015/05/22/politics/patriot-act-debate-explainer-nsa/index.html.

Complaint Against Faisal Shahzad. 2010. https://www.cbsnews.com/htdocs/pdf/shahzad.pdf.

DeRosa, Mary. 2005. "Lone Wolf Amendment." Blog. Patriot Debates. https://apps.americanbar.org/natsecurity/patriotdebates/lone-wolf.

Doyle, Charles. 2002. "The USA PATRIOT Act: A Legal Analysis." RL31377. Congressional Research Service.

FBI | Countering Violent Extremism. n.d. Accessed March 4, 2019. https://cve.fbi.gov/.

Fischer, Eric A. 2014. *Federal Laws Relating to Cybersecurity: Overview of Major Issues, Current Laws, and Proposed Legislation*. R42113. Washington, DC: Congressional Research Service.

Foreign Assistance Act of 1969. 1969. *22*. Vol. 2162.

Foreign Intelligence Surveillance Act of 1978. 1978. *50*. Vol. 1801.

Foreign Intelligence Surveillance Court | United States. n.d. United States Foreign Intelligence Surveillance Court. Accessed March 25, 2016. http://www.fisc.uscourts.gov/.

Foreign Terrorist Organizations. n.d. U.S. Department of State. Accessed April 1, 2019. http://www.state.gov/j/ct/rls/other/des/123085.htm.

Global Terrorism Database. 2018. National Consortium for the Study of Terrorism and Responses to Terrorism (START). http://www.start.umd.edu/gtd.

Gruenewald, Jeff, Steven Chermak, and Joshua D. Freilich. 2013. "Distinguishing 'Loner' Attacks from Other Domestic Extremist Violence: A Comparison of Far-Right Homicide Incident and Offender Characteristics." *Criminology & Public Policy* 12, no 1: 65–91. https://doi.org/10.1111/1745-9133.12008.

Hoffman, Bruce. 2016. "Lone Wolf Is Useless Analytical Category. Unabomber & Adam Lanza Were LWs. Inspired by Stated Terrorist Group Strategy Something Different." Microblog. *@hoffman_bruce* (blog). June 14, 2016. https://twitter.com/hoffman_bruce/status/742752855371616257?refsrc=email&s=11.

Hoffman, Grayson A. 2003. "Litigating Terrorism: The New FISA Regime, the Wall, and the Fourth Amendment." *The American Criminal Law Review* 40, no. 4: 1655–82.

Homegrown Violent Extremism. n.d. Audio. Federal Bureau of Investigation. Accessed October 28, 2016. https://www.fbi.gov/audio-repository/news-podcasts-thisweek-homegrown-violent-extremism.mp3/view.

International Security Assistance and Arms Export Control Act of 1976. 1976. *22*. Vol. 2151.

LaFree, Gary. 2013. "Lone-Offender Terrorists: Loner Attacks and Domestic Extremism." *Criminology & Public Policy* 12, no. 1: 59–62. https://doi.org/10.1111/1745-9133.12018.

Ludwick, Keith. 2016. "The Legend of the Lone Wolf: Categorizing Singular and Small Group Terrorism." Dissertation, Fairfax, Virginia: George Mason University.

Matz, Chris. 2008. "Libraries and the USA PATRIOT Act: Values in Conflict." *Journal of Library Administration* 47, no. 3–4.

McNally. 2015. *The State of Tennessee Senate Resolution 16*.

Miklaszewski, Jim, Erin McClam, and Courtney Kube. 2015. "Military Plans Increased Security at Recruiting Stations After Chattanooga Attack." NBC News. July 20, 2015. https://www.nbcnews.com/storyline/chattanooga-shooting/military-plans-increased-security-recruiting-stations-after-chattanooga-attack-n395266.

Municode Library. 2015. Oklahoma City, Oklahoma—Code of Ordinances. August 15, 2015. https://www.municode.com/library/ok/oklahoma_city/codes/code_of_ordinances.

National Commission on Terrorist Attacks upon the United States. ed. 2004. *The 9/11 Commission Report: Final Report of the National Commission on Terrorist Attacks upon the United States*. 1st ed. New York: Norton.

Pantucci, Raffaello. 2011. *A Typology of Lone Wolves: Preliminary Analysis of Lone Islamist Terrorists*. London: Kings College.

P.A.T.R.I.O.T. Act of 2001. 2001. *18*. Vol. 1.

Phillips, Peter J. 2011. "Lone Wolf Terrorism." *Peace Economics, Peace Science and Public Policy* 17 (March): 1–29. https://doi.org/10.2202/1554-8597.1207.

Sinai, Joshua. 2012. "Terrorism Bookshelf: Top 150 Books on Terrorism and Counter-Terrorism." *Perspectives on Terrorism* 6, no. 2. http://www.terrorismanalysts.com/pt/index.php/pot/article/view/sinai-terrorism-bookshelf.

"Special Report: A Review of the FBI's Handling of Intelligence Information Related to the September 11 Attacks (Full Report)." 2006. *Office of the Inspector General*. Washington, DC: Department of Justice.

Statement of David Kris, Assistant Attorney General Before the Committee on the Judiciary, United States Senate Entitled "Reauthorizing the USA Patriot Act: Ensuring Liberty and Security." 2009. Washington, DC.

"To Punish for Exerting Mutinous Influence upon Army and Navy." 1935. 74th Congress, 1st Session 1603. House of Representatives.

Tucker, Eric, and Michael Balsamo. 2018. "Attacks Renew Debate: Should US Have Domestic Terrorism Law?" AP News. October 30, 2018. https://www.apnews.com/dba83c3a4f11457f852af74cbea1b96e.

Turse, Nick. 2018. "Is the US Military Winning the War on Terror? Sure." *The Nation* (blog). September 4, 2018. https://www.thenation.com/article/is-the-us-military-winning-the-war-on-terror-sure/.

U.S. Senate: Watergate. 2014. United States Senate. June 2, 2014. http://www.senate.gov/artandhistory/history/common/investigations/Watergate.htm.

USA Patriot Improvement and Reauthorization Act of 2005. 2005. *18*. Vol. 1801.

Wildenthal, Bryan H. 2017. "Indian Sovereignty, General Federal Laws, and the Canons of Construction: An Overview and Update." *American Indian Law Journal* 6, no. 1: 98–173. https://doi.org/10.2139/ssrn.2987620.

Worth, Kaite. 2016. "Lone Wolf Attacks Are Becoming More Common—And More Deadly." FRONTLINE. July 14, 2016. http://www.pbs.org/wgbh/frontline/article/lone-wolf-attacks-are-becoming-more-common-and-more-deadly/.

Zelman, Joshua D. 2001. "Recent Development in International Law: Anti-Terrorism Legislation—Part One: An Overview." *Journal of Transnational Law & Policy* 11: 183–201.

Zimmerman, Katherine. 2018. "The Never-Ending War on Terror." *Foreign Affairs*, May 11, 2018. https://www.foreignaffairs.com/articles/2018-05-11/never-ending-war-terror.

Zimmermann, Doron, and Andreas Wenger, eds. 2007. *How States Fight Terrorism: Policy Dynamics in the West*. Boulder, CO: Lynne Rienner Publishers.

The Neglected Dimension of Ideology in Russia's Political Warfare Against the West

Armin Krishnan

East Carolina University, Greenville, NC, USA

Abstract

The Cold War was in its essence an ideological conflict where the political ideas of Western liberal democracy competed with communism. The New Cold War with Russia has claimed to be a purely geopolitical conflict or a conflict over the influence and control of world regions and resources. The dimension of ideology tends to be neglected in Western analyses of Russia's political warfare, which focus mostly on the propaganda and disinformation aspects. This article argues that Russia has developed a new ideology that again presents itself as a viable alternative social model to the West and it has an alternative vision of world order. Russia's new ideology contains aspects that make it appealing to larger societal groups in the West, both on the political right and on the political left. In order to win this new "war of ideas," the West must take Russian ideology seriously and must engage in this "war of ideas" as was done during the original Cold War.

Keywords: Russian ideology, political warfare, Eurasianism, propaganda

La dimensión descuidada de la ideología en la guerra política de Rusia contra Occidente

Armin Krishnan

East Carolina University, Greenville, NC, EE. UU.

Resumen

La Guerra Fría fue en esencia un conflicto ideológico donde las ideas políticas de la democracia liberal occidental competían con el comunismo. La Nueva Guerra Fría con Rusia ha afirmado ser un

doi: 10.18278/gsis.4.2.3

conflicto puramente geopolítico o un conflicto sobre la influencia y el control de las regiones y recursos mundiales. La dimensión de la ideología tiende a ser descuidada en los análisis occidentales de la guerra política de Rusia, que se centran principalmente en los aspectos de propaganda y desinformación. Este artículo argumenta que Rusia ha desarrollado una nueva ideología que nuevamente se presenta como un modelo social alternativo viable para Occidente y tiene una visión alternativa del orden mundial. La nueva ideología de Rusia contiene aspectos que la hacen atractiva para grupos sociales más grandes en Occidente, tanto en la derecha política como en la izquierda política. Para ganar esta nueva "guerra de ideas", Occidente debe tomar en serio la ideología rusa y debe participar en esta "guerra de ideas" como se hizo durante la Guerra Fría original.

Palabras clave: Ideología rusa, guerra política, eurasianismo, propaganda

俄罗斯对西方的政治战中被忽视的意识形态维度

Armin Krishnan

East Carolina University, Greenville, NC, USA

摘要

冷战的本质是由西方自由民主与共产主义各自政治观念产生的意识形态冲突。俄罗斯遭遇的新冷战宣称为一次纯粹的地缘政治冲突、或者说是一次有关控制和影响全球区域和资源的冲突。西方在分析俄罗斯政治战时倾向于忽略其意识形态的维度，而基本上聚焦于政治宣传和假新闻等方面。本文主张，俄罗斯已提出一种新意识形态，这种思想再次将俄罗斯展现为一个可行的、能代替西方的社会模式，并且它对世界秩序有另一种看法。俄罗斯的新意识形态所具备的某些方面能让其对更大的西方社会团体具有吸引力，包括政治右倾派和左倾派。为了在这场新的"观念战"中获胜，西方必须严肃对待俄罗斯意识形态，必须参与"观念战"，就像在第一次冷战期间一样。

关键词：俄罗斯意识形态，政治战，欧亚主义，政治宣传

Introduction

As late as 2014, President Obama claimed that the current tensions with Russia would not be "another Cold War that we're entering into...After all, unlike the Soviet Union, Russia leads no bloc of nations, no global ideology. The United States and NATO do not seek any conflict with Russia. In fact, for more than 60 years we have come together in NATO not to claim other lands but to keep nations free" (Miller 2014). Now, 5 years later, there is a growing realization by Western foreign policy elites that the conflict with Russia is of a more permanent nature and has great similarities to the original Cold War. In 2018, the President of the influential Council on Foreign Relations, Richard Haas, acknowledged "[a] quarter-century after the end of the Cold War, we unexpectedly find ourselves in a second one" (Haas 2018). Many other analysts and journalists have similarly claimed that there is a return of a Cold War relationship between the West and Russia (Conradi 2017; Legvold 2016; Lucas 2008). Russia is increasingly seen as a revanchist and revisionist power that seeks to challenge U.S. hegemony by waging a political war against the West (Nance 2018).

However, unlike the original Cold War, which is widely understood as an ideological conflict that pitted liberal democracy against communism, it has been claimed that the new Cold War completely lacks any ideological dimension. According to Peter Pomerantsev, the Kremlin has no distinctive ideology or rather that its ideology would be the belief that "there is no truth." In Pomerantsev's view, Russia discarded any attempt of having a consistent ideology in favor of the aim to "own all forms of political discourse" by "climb[ing] inside all ideologies and movements, exploiting and rendering them absurd" (Pomerantsev 2015, 67). Similarly, the *Washington Post* claimed: "The rule-by-fear is Soviet, but this time there is no ideology—only a noxious mixture of personal aggrandizement, xenophobia, homophobia and primitive anti-Americanism" (Hiatt 2013). Analyst Benn Steil has suggested that "Russia's conflict with the West is about geography, not ideology," as Russia would merely try to preserve its sphere of influence against NATO encroachment (Steil 2018).

In contrast, other Russia specialists have claimed that Russia has simply turned fascist, which has been blamed for the current crisis in relations with the West. For example, historian Timothy Snyder argued that "[f]ascist ideas burst into the Russian public sphere during the Obama administration's attempt to 'reset' relations with the Russian Federation" and that "[t]he dramatic change in Russia's orientation bore no relation to any new unfriendly action from the outside" (Snyder 2018, 91). The main argument for describing Russian ideology as fascist is the fact that the Russian state is both nationalist and authoritarian and that there is a cult of personality with respect to Putin himself (Snegovaya 2017, 43). In particular, some analysts have noted the surprising political revival of the Russian far-right or "fascist" philosopher Ivan Ilyin in Putin's Russia in recent years (Laqueur 2015, 177).

This article instead argues that there is a new Cold War, that Russia has a somewhat consistent ideology, and that it has to be taken seriously as an alternative model of society and an alternative vision of world order. The new ideology underlies the new Cold War and is used offensively in Russia's political warfare against the West. But it would be an enormous simplification to call Russian ideology fascist or to assume that Russia new ideology would only be attractive to fascists and social misfits. Unless this new Cold War is properly understood as a "war of ideas" and unless Russia's ideology is adequately addressed in the Western counterpropaganda, the West will not be able to persevere, as Western governments will be increasingly outmaneuvered by Russian information warfare. The article will sketch the key elements of Russian ideology and explain how it connects to political movements in the West.

Ideology and Political Warfare

The term "ideology" was coined by Count Antoine Destutt de Tracy in the late eighteenth century. He understood ideology as scientific "system of truths closely tied together" about the social and political order that is empirically verifiable (Steger 2008, 19–24). The concept has been reinterpreted by many philosophers to mean many things. Michael Freeden proposed a definition of ideology as "a set of ideas, beliefs, opinions, and values that (1) exhibit a recurring pattern (2) are held by significant groups (3) compete over providing and controlling plans public policy (4) do so with the aim of justifying, contesting or changing the social and political arrangements and processes of a political community" (Freeden 2003, 32). The main ideologies of the twentieth century were liberalism, communism, and fascism. After the defeat of fascism at the end of World War II, the main ideological competition that remained was between American liberalism and Russian/Soviet communism. According to Manfred B. Steger, "[i]deology [in the Cold War] functioned as a critical factor in an epic contest that was just as much about showcasing the global applicability of liberal or communist ideas and values as it was about economics or geopolitics" (Steger 2008, 135).

The Cold War was fundamentally ideological in nature and was fought primarily by political means rather than by military engagements, which has been described as "political warfare." The term "political warfare" originates from the Second World War era, although the practice is much older (Hancock 2018, 12–13). It relates primarily to the use of propaganda, disinformation, and psyops aimed at weakening an enemy society, enemy forces, an enemy's ability to form and maintain alliances, and enemy decision making. The term was popularized by the American diplomat George Kennan, who suggested that it would include "overt actions as political alliances, economic measures (as ERP), and 'white' propaganda to such covert operations as clandestine support to 'friendly' foreign elements, 'black' psychological warfare and even the encouragement of underground resistance in hostile states" (Kennan 1948). According to Kintner and Kornfelder, "political warfare seeks to induce the desire for surrender—in opposing peoples. At the strategic level, political warfare seeks to corrode the entire moral, political, and economic infrastructure of a nation, particularly by affecting governmental decisions…Political warfare aims to weaken, if not destroy, the enemy by use of diplomatic proposals, economic sorties, propaganda and misinformation, provocation, intimidation, sabotage, terrorism, and by driving a wedge between the main enemy and his allies" (Kintner and Kornfelder 1962, XIII).

The way political warfare was practiced by the Soviet Union was a program of ideological subversion as explained by the Soviet defector Yuri Bezmenov. According to Bezmenov, the ultimate goal of Soviet ideological subversion was "to slowly replace the free market capitalist society, with its individual freedoms in economic and socio-political spheres of life—with a carbon-copy of the 'most progressive system, and eventually merge into [a] one world-wide system ruled by a benevolent bureaucracy that they call Socialism'" (Schuman 1985, 5). This approach of subversion is also known as "active measures" in Soviet terminology. According to Richard Shultz and Roy Godson, active measures "are employed to influence the policies of other governments, undermine confidence in the leaders and institutions of these states, disrupt the relations between various nations, and discredit and weaken major opponents. This frequently involves attempts to deceive the target, and to distort the target's perceptions of reality" (Shultz and Godson 1984, 16). In practice, active measures amounts to political covert action, namely the creation and use of political front organizations, the recruitment of "agents of influence," the financing of certain sympathetic political parties and individuals, the dissemination of propaganda at a grassroots level, and the use of fabrication, forgeries, and disinformation designed to increase public suspicions and distrust of their governments.

Active measures are long-term campaigns that are expected to last years, if not decades, in order to produce a cumulative corrosive effect on targeted societies. Active measures being "a major component of Soviet foreign policy, were incredibly well-resourced" (Abrams 2016, 7–8). The Soviets used their political front organizations and propaganda to undermine Western values and morality in order to pave the way for socialism. Similarly, the West was also "weaponizing" information and culture to push back against harmful communist influences in Western societies. This ideological battle "was fought in classrooms and on college campuses, in journals and books, and in radio broadcasts, television programs, and the silver screen, and, of course, in the courts, and it involved the use of a plethora of catchwords and images" (Echevarria 2008, 15). It was a total cultural and political battle for the minds of men, if not their souls. Then the Cold War ended and seemingly with it the ideological struggle.

It has been subsequently claimed that ideology and other "grand narratives" are effectively dead after the collapse of communism and the rise of the postmodern values of diversity, relativity, rejection of materialism and authority, and permissiveness ("anything goes"). Francis Fukuyama declared *The End of History* in his 1992 book with the argument that liberalism has eventually defeated the only remaining competing ideology of communism, leaving the world with only one ideal as to how society should be organized (as a liberal democracy with a free-market economy). Only after 9/11 and with the beginning of the War on Terror, there was realization that the West was in another "war of ideas" or a new ideological struggle with radical Islam. President George W. Bush declared in 2006: "The war against this enemy is more than a military conflict. It is the decisive ideological struggle of the 21st century, and the calling of our generation" (Bush 2006). This has led U.S. strategists to conclude that a victory in the War on Terror would require the West to win the war of ideas as well as the physical war on the battlefield (Echevarria 2008, 23–24). Now that the War on Terror seems to be winding down, Russia has reemerged as a new threat to the West and to the world order that was formed after 1945. Russia has again turned to ideology and political warfare to achieve its political objectives.

The Origins of the New Cold War

Despite initial hopes that the nature of the relationship between Russia and the West could be transformed and the Cold War could be ended for good after the collapse of the Soviet Union, there was never enough trust between both sides to make it happen. On the Russian side, there was early on the firm belief that the Soviet Union and communist bloc did not simply collapse because of internal reasons but rather collapsed because of covert Western influence and manipulation (Legvold 2016, 83–84). Mikhail Gorbachev and Boris Yeltsin were perceived by the former Soviet elites to have been traitors to their nation for bowing to Western demands. Furthermore, the West was seen as taking advantage of a massively weakened Russia when it moved to expand NATO and attack its ally Serbia in 1999.

Russian elites disagreed with the tendency of Westernization under Yeltsin and they became increasingly concerned about the ideological threat posed by the West to Russian sovereignty and Russian national security. According to a popular Russian conspiracy theory, CIA Director Allan Dulles had devised a diabolic plan in the 1940s to completely demoralize Soviet society. A document attributed to Dulles states: "After sowing chaos there [in Russia], we shall imperceptibly replace their values by stealth with false ones and shall force them to believe in these false values" (Shiner 2018). In this view, Russia has been under an informational attack from the West since the 1950s, which ultimately caused the Soviet collapse and which aims to destroy Russian national identity and marginalize Russia as country by breaking it into pieces. Russian theorists like to point at the existence of a "noosphere" or "mental sphere" that shapes the thinking and behavior of a people. According to the Russian theorist Vladimir Karyakin, "[t]he mental sphere, a people's identity, and its national and cultural identity have already become battlegrounds. The first step in this direction is the discrediting of and then the destruction of a nation's values. And in order for external aggression to be perceived painlessly to the mass consciousness, it must be perceived as movement along the path of progress" (quoted from Blank 2013, 33).

In particular, Russian national security elites became concerned about Western influences on the Russian youth through Western media and the Internet, the introduction of damaging values and attitudes that undermine Russian identity, and the threat of political and societal destabilization through what has been later termed "color revolutions" or popular uprisings at Russia's periphery that overthrow governments and replace them with pro-Western governments (Karyakin 2013).

These are not fringe ideas but they are actually echoed in official Russian policy documents. For example, the Russian *Information Security Doctrine* from 2000 already contained ideas regarding the new informational and hybrid war threat (also emphasized in Russia's 2014 military doctrine). The Information Security Doctrine argues that information has become a key factor in societal development and that "the national security of the Russian Federation substantially depends on the level of information security." As a major threat, it lists "illegal use of special means of influence on individual, group and public consciousness"; the "ousting of Russian news agencies and media from the national information market, and an increase in dependence of the spiritual, economic and political areas of public life in Russia on foreign information entities"; and the "depreciation of spiritual values, the propaganda of specimens of mass culture based on the cult of violence or on spiritual and moral values contrary to the values adopted in Russian society" (Russian Federation 2000).

Stephen Blank has characterized Russian paranoia over Western influence on Russian society as a "domestic counterinsurgency" since it would lend legitimacy to employing information warfare and other measures for internal repression (Blank 2014). The Russian government has apparently decided that the best way of fighting harmful Western influence on Russian society is by waging information warfare against the West. In fact, Russian active measures campaigns in the West have continued throughout the post-Cold War era and may have intensified over the last decade (Abrams 2016, 17–18). After the election of Vladimir Putin as president in 2000, Russia has strengthened its "soft power" by establishing the foreign language TV news channel *Russia Today*, as well as setting up Internet news websites like *Sputnik News* (formerly RIA Novosti) and *Ruptly*. At one point, the budget for *Russia Today* exceeded $300 million and it represents a substantial investment in strengthening the Russian media presence abroad (Diamond, Plattner, and Walker 2016, 51). *Russia Today* (rebranded as RT in 2009) has 21 offices in 16 countries, including in London, Washington, DC, Berlin, and Paris broadcasting in English, Spanish, French, German, and Arabic. RT set up a video on demand web service with the name *Ruptly* in 2013. Altogether RT reaches 700 million people worldwide and it has close to 1.2 billion views of its news clips on Youtube and is, as a result, surpassing CNN as an international news source (Spiegel Staff 2014). Putin also created the Valdai Forum, which is a Russian version of the Davos meeting and which aims to expand Russian influence by courting Western intelligentsia and politicians. Russia also followed the Western example of establishing and funding various NGOs that promote Russian culture and values. The purpose of the *Russky Mir* Foundation, created in 2007, was to engage with the Russian diaspora abroad, "predicated on the idea that Russian speakers across the world make up one unified civilization" (Pomerantsev and Weiss 2014, 19). In 2009, 14 of these NGOs were organized in the Coordinating Council of Russian Compatriots (Perry 2015). The idea is to create relationships with ethnic Russians in the near-abroad, so that they can be mobilized in support of Russian foreign policy, as was done in Estonia, Georgia, and more recently in Ukraine.

The Search for a New Russian Ideology

Russia has no official or explicit state ideology because it is prohibited by the Russian constitution, which states under Article 13.2 "[n]o ideology may be established as state or obligatory one." This provision was included to prevent a return of totalitarianism in Russia. Besides, Marxism was in a severe crisis in the early 1990s, leaving Russia with no ideological orientation of its own. Since there was the widespread perception that Russian identity was under attack, a new ideology had to be developed that could provide a new vision for Russia's national identity and for Russia's role in the world. This ideological gap was soon to be filled by a hodgepodge of ideas and interconnected concepts, many of which were borrowed from Western thinkers, that are now known as "Neo-Eurasianism." This school of thought is derived from Eurasianism, which is a political ideology that had its origins in the 1920s. It was originally conceived as an alternative to both communism and Western parliamentary democracy by Russian émigrés living in the West. According to Marlene Laruelle, "Eurasianism was thus born in the context of a crisis [the Bolshevik takeover of Russia], in an atmosphere of eschatological expectations: Its proponents had the feeling of standing at a turning point in human history. Their attempts to theorize these expectations made them look toward the future" (Laruelle 2012, 19). Its main thesis is that Russian civilization does not belong to Europe or Asia but would be distinctive from both of these civilizations.

The most well-known Russian traditionalist thinker developing contemporary or Neo-Eurasianism is Alexander Dugin. Dugin is a former chair of the sociology department of Moscow State University. He is also a former member of the National Bolshevik Party, the founder of the Eurasia Party (in 2002), and the founder of the Eurasian Youth Movement (in 2004). Dugin is one of the best-known and most controversial Russian political philosophers. Anton Barbashin and Hannah Thoburn called Dugin "Putin's Brain" and suggested that Dugin's ultra-conservative ideas have become very popular in Putin's Russia (Barbashin and Thoburn 2014). Similarly, John Dunlop claimed that "[b]y summer 2001, Aleksandr Dugin, a neo-fascist ideologue, had managed to approach the center of power in Moscow, having formed close ties with elements in the presidential administration, the secret services, the Russian military, and the leadership of the state Duma" (Dunlop 2004). Although Dugin was never the official ideologue of the Kremlin, "[t]he worldview he advocates has become part of mainstream thinking, both in the Russian political establishment and among the general public" (Ligerant 2009). Dugin may have lost his influence on the Kremlin since his outrageous comments regarding the Ukraine crisis of 2014, which had cost him his job at Moscow State University, but Dugin's ideas remain influential.

Dugin's philosophy is largely a repudiation of Western liberalism and an attempt to provide Russia with a national identity and a historical mission after the collapse of the Soviet Union. Dugin's Neo-Eurasianism is a strange mixture of various "intellectual traditions, such as theories of conservative revolution, the German geopolitics of the 1920s and 1930s, René Guénon's Traditionalism and the Western New Right" (Laruelle 2006, 5). Although Dugin is a critic of Marxism and may be considered a dissident during the Soviet period, he is a firm collectivist with socialist leanings. "Inspired by philosophers closely associated with fascism and Nazism, Dugin is an outspoken critic of capitalism, liberal democracy, and the bourgeois social order," suggests Yigal Ligerant (Ligerant 2009). It is not easy to sum up his political theory, but there are some core ideas that have seemingly influenced the rhetoric and foreign policy of the Russian government.

Geopolitics and World Order

Russian ideology is often mischaracterized as "geopolitics" because Russian theorists and strategists frequently invoke the term when explaining the conflict with the West. Indeed, there was a great revival of geopolitical thought in the Russia of the 1990s. However, geopolitics is a fuzzy concept. It largely explores the relationship between geography and politics, often in relation to natural borders, "large spaces," trade routes, influence over other states, and access to natural resources. Geopolitics emerged as an intellectual school of thought in the early twentieth century and has been associated with both imperialism and Nazism. The main thinkers of geopolitics were Halford Mackinder, Thomas Mahan, Rudolf Kjellen, Frederick Ratzel, and Karl Haushofer.

Dugin borrowed from these thinkers when he published his first major work *The Foundations of Geopolitics* in 1997. The book was used as a textbook in many Russian educational institutions, including (presumably) the Russian General Staff Academy (Dunlop 2004). Dugin claims, in line with geopolitical thinkers of the early twentieth century, that world history is driven by the perpetual conflict between land powers (Eurasianists) and sea powers (Atlanticists), which will last until one of them is destroyed (Barbashin and Thoburn 2014). In reference to Mackinder's "heartland theory," which suggested that control of Eastern Europe would hold the key to the rule of the world, Zbigniew Brzezinski argued in his 1996 book *The Grand Chess Board* that America shall prevent a unification of the Eurasian landmass by fostering independent states in Central Asia in order to preserve U.S. hegemony. Dugin therefore argues that Russia must unite Eurasia to defeat the main enemy of Atlanticism. In particular, he proposes three political axes to achieve this goal: a Moscow–Berlin axis, a Moscow–Tokyo axis, and a Moscow–Teheran axis. In deviation from classical geopolitical thought, Dugin defined Russia as a civilization, picking up Samuel P. Huntington's theory of a "clash of civilizations." Similar to Huntington, who identified Russia as the core state of an "Orthodox civilization," Dugin argues that Russia has no choice but to be an empire as it is not a racially uniform state. "A repudiation of the empire-building function…would signify the end of the Russian people as a historical reality, as a civilizational phenomenon. Such a repudiation would be tantamount to Russian suicide" (quoted from Dunlop 2004).

Dugin sees the Eurasianist project as an alternative to "Americanisation, Westernisation, and globalization" and proposes a "model of world order based on the paradigm of unique civilisations and Great powers. It presupposes the creation of different transnational political, strategic, and economic entities united regionally by the community of common geographic areas and shared values" (Dugin 2012, 80–81). American hegemony or Atlanticism would be replaced by concepts such as multi-polarity, Great Spaces, and Great Powers. Eurasianism or the unification of the Eurasian space would be Russia's destiny and historical mission. In terms of geography, Eurasia would include the countries of Central and Eastern Europe, Manchuria, Xinjiang, Tibet, Mongolia, and the Orthodox parts of the Balkans (van Herpen 2015, 76). It could stretch "from Dublin to Vladivostok" (Bassin 2007, 293). In effect, the geographical space of Eurasia that Russia seeks to dominate would be significantly bigger than the Soviet Union, which has resulted in fears by Eastern European states of Russia's renewed imperial ambitions.

Putin seems to have followed Dugin's argument when he declared at the Munich Security Conference in 2007: "I consider that the unipolar model is not only unacceptable but also impossible in today's world. And this is not only because if there was individual leadership in today's—and precisely in today's—world, then the military, political and economic resources would not suffice. What is even more important is that the model itself is flawed because at its basis there is and can be no moral foundations for modern civilisation" (Putin 2007). Putin pointed out that the U.S. imposes "economic, political, cultural and educational policies" on other nations, asking rhetorically: "Who likes this?"

Russia is now fighting for a multi-polar world that it seeks to create by unifying Eurasia and by putting into place alternative global institutions. Russia spearheaded the BRICS summits since 2009, which is an annual economic forum and international organization, bringing together Brazil, Russia, India, China, and South Africa. In 2015, Russia established the Eurasian Economic Union, which consists of Russia, Armenia, Belarus, Kazakhstan, and Kyrgyzstan, "which has Dugin's intellectual fingerprints all over it" and which seems to be the main instrument for implementing its vision of a multipolar world (Tolstoy and McCaffrey 2015). Although Dugin suggested to seek a closer partnership with Japan since China was a geopolitical rival and a base for Atlanticism, the Kremlin has opted for closer security and trade relations with China (Van Herpen 2016, 193).

Putin has frequently referred to Russia as a "state-civilization" and he has also rejected the notion that Russia would be in an ideological conflict with West, arguing that there were "culture-logical" or civilizational differences shaping relations between Russia and the West (Tsyngankov 2016, 146). By redefining Russia as being not only a state, but a state-civilization Russian ideology can solve several problems at once. Russia as a state compromised of multiple ethnicities, religious groups, and nationalities cannot invoke Russian ethnic nationalism without the risk of the country falling apart. It is the goal of Russian elites to maintain the Russian state in its present geography while expanding its influence by setting itself up as the leader of an Orthodox civilization that comprises many states, mostly in the former Soviet space. This way Russia can violate the sovereignty of countries in its civilizational space while insisting on its own sovereignty being sacrosanct. It has been pointed out by Steven Fish that the Russian leadership is deliberately ethnically very inclusive and that Putin has made strides to appease ethnic and religious minorities (Fish 2017, 66–67). Russian nationalism has steered clear of ethnic chauvinism and is interpreted as "patriotism" or a commitment to the authority and well-being of the Russian state. As a result, it would be inaccurate to describe Russian ideology as fascist since it clearly lacks the racist, if not anti-semitic, aspects of fascism and national socialism. At the same time, there is some apparent anti-semitism that can be found in Neo-Eurasianism and its proponents (Snyder 2018, 91–93).

In terms of ideology, Russia's message of multi-polarity, anti-globalization, and rejection of American hegemony is appealing to many people around the world. Russian propaganda has targeted Western interventionism as imperialism and has pointed out that U.S. interventions in countries like Iraq, Afghanistan, Libya, and Syria have been highly destabilizing for the international system. Even many U.S. allies have become suspicious of U.S. intentions in the aftermath of the Iraq War, the drone wars, and the Snowden revelations. There is a growing nationalist sentiment in many Western countries that Russia seems to be nurturing in the hope of undermining Atlanticism. Russia has supported Western politicians and political parties that are skeptical of global and regional institutions such as the UN, NATO, the IMF, and the EU, regardless whether they are on the political right or left. One could argue that this could be part of Dugin's plan of the "Finlandization of all Europe," meaning to bring Europe under Russian influence by undermining U.S. influence (Dunlop 2004).

Nationalist movements in Europe and North America have greatly benefitted from the migration crisis of 2015, which brought over a million migrants from the Middle East and North Africa to Europe. Russian propaganda has emphasized the dangers of migrants from Islamic states to the security and culture of Europe, which is in line with its concept of a civilizational conflict and which resonated well with larger portions of populations in Western Europe. Arguably, the migration crisis has enabled the rise of populist/nationalist leaders and of Brexit. The Russian government and Dugin have been aggressively establishing contacts with far-right groups and far-right leaders in Europe from France's National Front to Austria's FPÖ, Britain's UKIP, Denmark's Danish People Party, Hungary's Jobbik, and Greece's Golden Dawn Party (Polyakova 2014, 36–37). Relations between Europe and the United States have worsened substantially after the entirely unexpected election of Donald Trump in 2016. President Trump has antagonized European leaders by ending negotiations for TTIP, by demanding more NATO burden-sharing, and by unilaterally cancelling the 2015 Iranian nuclear deal, among other things.

Recently, Russian foreign minister Sergey Lavrov claimed that globalism "is losing its attractiveness and is no more viewed as a perfect model for all. Moreover, many people in the very western countries are skeptical about it" (Lavrov 2019). A recent opinion poll by IPSOS and King's College London, covering 17,000 adults in 24 countries, indicates that Russia is not far behind the United States in terms of its global reputation. While 18% see the United States as an influence for good and 22% see it as an influence for bad in the world, 13% see Russia as an influence for good, and 25% see it as an influence for bad in the world (IPSOS 2019). Russia's ideological vision of a multipolar world that would end U.S. global hegemony and would potentially push the United States out of Europe has become acceptable to a greater number of people in the West, both on the political right and on the left, which does not suggest that Russia would be promoting fascism or that only fascists would be receptive to Russian ideology or Russian viewpoints.

Collectivism and State Capitalism

Russian ideology rejects pluralism and liberal democracy and instead offers authoritarian collectivism and state capitalism as an alternative societal model. Collectivism means that the needs of the collective are more important than the needs or rights of individuals. Dugin states it bluntly as: "[t]he nation is everything; the individual is nothing" (Dunlop 2004). A modern form of collectivism is called statism, which is the idea that the state should be in control of the economy and shape society through extensive social policies. Collectivism is inherently authoritarian as there has to be a central authority that determines what is good for the collective and that can impose decisions on the collective without the need for constant political negotiation and without the constant balancing of diverse interests as is the case in a pluralist society.

Marxist socialism was based on the idea that it is the state that determines what people need and then provides it to them for free in return for their labor and other cooperation. Although Russia is no longer a socialist country in the strict sense of the state completely controlling the economy and providing extensive social services to the population, it is not a capitalist country either since Russia lacks strong property rights (Kolesnikov and Vertov 2019). Key aspects of the Russian economy are state-controlled, most importantly the oil and gas industries. Where the state has no direct control, it can always threaten nationalization as in the case of Yukos.

The Russian government justifies the lack of political freedom and real capitalism through a combination of populism and a proclaimed need to defend the Russian state against subversive influences from the West. Political opposition is tolerated as long as it is weak and, or under government control (Shevtsova 2016, 49). The Russian government has redefined its peculiar form of democracy as a "sovereign democracy," which means that the Russian state can itself determine the standards by which it is a "democracy" (Glazunova 2018, 62). Attempts to make Russia more democratic can be declared to be a harmful foreign influence aimed at undermining the Russian state and destroying it as a Great Power and a major influence on world history. Putin continues to be a popular leader in Russia, receiving a boost in his popularity after the annexation of Crimea (Fish 2017, 66).

Russia's closest foreign partners are mostly collectivist/socialist authoritarian states, including Belarus, Armenia, Kazakhstan, Kyrgyzstan, Tajikistan, China, Iran, Syria, Cuba, Vietnam, North Korea, and Venezuela, and to some extent Turkey and India. Russia wants to lead an anti-Western alliance of socialist/authoritarian states in a new multipolar world (Chandler 2008, 101–40). Russian propaganda attacks liberal democracy by pointing out the corruption and ineptness of many Western democracies. "The goal is to prove that the West is just as bad as the regimes that the West is trying to criticize" (Shevtsova 2016, 51). This plays into the hands of the populist movements in Europe and North America, which have emerged because of a general disappointment with the democratic political process and the perceived unwillingness of the political class to protect the interests and address the concerns of the common people. Many of the new populist movements in Europe and North America have a collectivist statist ideology, which favors massive government intervention in society to address certain social issues such as immigration, poverty, inequality, social justice, or environmental degradation.

It has been claimed that "socialism is re-emerging in both Europe and North America," although it is different from the socialism of the twentieth century (Pejovich 2018, 118). The new "liberal socialism" is different from the old socialism in the sense that it supports free elections and individual property rights (Pejovich 2018, 120). America has now over 23 presidential candidates for the 2020 election, many of whom even self-identify as being "democratic socialists" and who promise a substantial redistribution of wealth through "higher taxes for the rich" and extensive social programs. The problem is that "liberal socialism" is still an inherently collectivist and statist ideology in the sense that it declares the state as the authority to determine how wealth is to be appropriately distributed and how society should be regulated by the government in the pursuit of an assumed common good. This would not be substantially different to the Russian and Chinese model, except that their governments are much more corrupt. The West does not only seem to be unable to offer a clear alternative to the state capitalism of the authoritarian collectivist countries, most notably Russia and China, but increasingly moves in the very same direction of an ever-more regulated economy and society with less freedom. This creates a credibility gap when the Russian model is criticized.

Conservativism and Russian Orthodox Faith

Russian ideology is at its core conservative (Glazunova 2018, 59). However, Russian conservativism is not the same as liberal conservativism: it is traditionalism, which rejects modernity in favor of religious spirituality. According to Yigal Ligerant, Dugin bizarrely "has called for the restoration of Russia's medieval social hierarchy, with an aristocratic ruling class under religious patronage" Ligerant 2009). Dugin claims that liberalism has an inherent logic that is self-destructive: "having decided to liberate itself from everything that keeps man in check, the man of the modern era reached his logical apogee: before our eyes he is liberated from himself. The logic of world liberalism pulls us into the abyss of postmodern dissolution and virtuality" (Dugin 2014a, 2014b, 154). Liberalism, it is claimed, separated state and church and man and God, which gave rise to perversion and wickedness. Liberalism, so Dugin, would be not just an "accidental heresy," but is rather "an absolute evil" (Dugin 2014a, 2014b, 154–55).

The alternative foundation for social morality other than philosophical liberalism would be the Russian Orthodox faith, which has experienced an astonishing revival in contemporary Russia. Dugin even argued for the "total churchification" of Russia for "the Russian nation to become viewed simply as 'the church'" (Dunlop 2004). For Dugin, the geopolitical struggle of Antlanticists versus Eurasianists is a battle between good and evil with the ultimate goal of bringing about the end times or God's final judgment of humanity (Dugin 2012, 183). "The American Empire should be destroyed. And at one point it will be…Spiritually, globalization is the creation of a grand parody, the kingdom of the Antichrist. And the United States is the centre of its expansion" (Dugin 2014a, 2014b, 193).

Russian society has always been deeply religious and religion was always used instrumentally by the rulers of Russia from the Tsars to the Soviet state (Van Herpen 2014, 129). The Orthodox faith does not support the notion of a separation of state and church and the Orthodox Church has traditionally served the rulers of Russia. The Bolsheviks tried hard to eliminate religion in Russia since it had no place in Marxism–Leninism. Marx was an atheist and materialist, who believed that history is driven only by material conditions and not by divine design or intervention. But not even Stalin's brutal suppression of religion could destroy the Orthodox Church in Russia. As a result, the Soviet state effectively took control of the Orthodox Church to make sure that their teachings would not get in the way of Soviet rule. The KGB not only infiltrated the Orthodox Church early on, but even used the church for their global propaganda efforts. They allowed it to join the World Council of Churches so that they could assist in the ideological subversion of other societies, primarily in the Third World by making Orthodox priests denounce "colonialism" (Andrew 2000, 635–37).

The 1990s were a period of a decline of religion in Russia but since Putin has assumed power, there has been a notable revival of religion and spirituality, which is no coincidence. Putin was not openly religious before 1993 and his change to a believer seems to have been motivated by a desire to advance his political career. When he became president, he made it a point to visit Orthodox churches and monasteries and to publicly display his faith (Van Herpen 2014, 131). He has a chapel in his residence for his spiritual needs and there are indications that his religiosity is more than just public pretense (Adamsky 2019, 93). Putin's government has sponsored the building of countless churches across Russia and has promoted the Orthodox Church and Russian culture abroad. This has to be understood to be not just as a charitable or purely cultural activity, but as an important part of the defense of Russia and of the Kremlin's concept of "spiritual security" (Adamsky 2019, 86). Most notably, the Kremlin has been keen to restore morality in Russia. Putin has pushed back against homosexuality, LGBT rights, abortion, pornography, feminism, and transgenderism in line with Orthodox values. The Russian government now also emphasizes the idea that Russia is the "third Rome" or the legitimate successor of the Roman empire after the fall of Constantinople in 1453. In other words, Russia portrays itself as the "last bastion of conservative values, whereas Christianity perished under the onslaught of immoral liberal ideas" (Galstyan 2016). Russian propaganda has thus continued the Soviet practice of chastising Western decadence and racism, which is a message to which many conservatives in the West are receptive.

The conservativism in Russia's ideology serves a variety of purposes: it supports the continuation of the existing social order in Russia by providing legitimacy to the state to minimize Western influence under the guise of protecting societal morality and Russian identity, it turns Putin into a leader of a spiritual awakening in the world that makes him popular with many conservative Christians in the West, it enables Russia to reach out to and influence Orthodox Christians outside of Russia and mobilize them in support of the Kremlin's foreign policy, and it provides further ideological justification for pushing back Westernization and Western liberalism in non-Western societies.

Russia ideology is also conservative in the literal sense: it seeks to preserve the world as it is and to push back against any international or cultural progressivism. Steven Fish has pointed out that "Putinism broadly prioritizes the maintenance of the status quo while evincing hostility toward potential sources of instability" (Fish 2017, 61). An ideological downside of this conservative stance is that it conflicts with a more revisionist or reformist foreign policy agenda. Russia must present its foreign policy in terms of preserving a status quo and the traditional order of things. It cannot openly advocate revolutionary change or social reform without contradicting its conservative stance. A Russia must be a champion of the status quo and it must remain committed to Christian Orthodox values, which also means that it must continue to oppose harmful American liberalism in world in something like a crusade for Orthodox Christendom.

Conclusion

John Schindler has argued that "Vladimir Putin and his regime possess an official ideology, as they state plainly" and that it would be important to pay attention to it and to take it seriously (Schindler 2018). The current conflict is not just about geopolitics: it is also about values, societal models, and visions of world order. This fundamental conflict cannot be resolved by making geopolitical concessions or by hoping for a new leadership in Moscow. It is bound to be a conflict that will last a long time. Russia has made it clear what stands for, namely Orthodox conservative values, benevolent authoritarianism and state capitalism, and a multipolar world in which Great Powers would balance each other. The actual reality behind this ideology is certainly different than its rhetoric, but what matters more is the attractiveness of ideas and the persuasiveness of narratives.

The West is in a great disadvantage in the new ideological conflict. First of all, it has yet to acknowledge that it is in competition with a new ideology. Secondly, its own ideology of liberalism seems exhausted and is no longer without a viable alternative. It is increasingly unclear for what exact values the West stands for as liberal progressivism is trapped in its own contradictions and absurdities, as pointed out by its vocal critics. Western liberal democracies are in a crisis as popular discontent with politics and institutions is growing. The Western economic model is no longer looking as strong as it was prior to the 2008 financial crisis, allowing a state-run economy like China to eventually surpass the U.S. economy in size. American hegemony is a mere shadow of its former self, as even some of the closest allies are no longer very responsive to U.S. demands. As John Schindler has argued, it is therefore "unwise to simply dismiss Westerners who have some sympathy for Putinism…as mere misogynists, racists and homophobes—that is, obvious retrogrades whom polite and decent people need not worry about" (Schindler 2018).

This means that the West has to either return to its ideological roots and virtues or it must devise a new ideology that is more attractive to more people in terms of its values, societal model, and vision of world order than its current form of progressivism and globalism. People in the West need to be educated about Russian ideology and warned about its totalitarian leanings. The societal consensus regarding liberal democracy and free-market capitalism must be strengthened again. Western political warfare has to fight back the way it did during the Cold War: it must win the war of ideas by demonstrating that it still has a better societal model, namely one that is based on individual freedom, free-market capitalism, and progress. Otherwise, people around the world will be left with only a single ideological choice: the choice to embrace state-imposed morality and statist collectivism in a world that is largely run by authoritarians.

References

Abrams, Steve. 2016. "Beyond Propaganda: Soviet Active Measures in Putin's Russia." *Connections* 15, no. (1): 5–31.

Andrew, Christopher. 2000. *The Mithrokin Archive: The KGB in Europe and the West*. London: Penguin Books.

Barbashin, Anton, and Hannah Thoburn. 2014. "Putin's Brain: Alexander Dugin and the Philosophy Behind Russia's Invasion of Crimea." *Foreign Affairs*, March 31.

Barbashin, Anton, and Hannah Thoburn. 2015. "Putin's Philosopher: Ivan Ilyin and the Ideology of Moscow's Rule." *Foreign Affairs*, September 20.

Bassin, Mark. 2007. "Eurasianism 'Classical' and 'Neo.'" In *Beyond the Empire: Images of Russia in the Eurasian Cultural Context*, edited by Tetsuo Mochizuki. Sapporo: Slavic Research Center.

Blank, Stephen. 2013. "Russian Information Warfare as Domestic Counterinsurgency." *American Foreign Policy Interests* 35, no. (1): 31–44.

Bush, George W. 2006. "Address to the Nation on the Fifth Anniversary of 9/11." Selected Speeches of President George W. Bush (2001–2008). https://georgewbush-whitehouse.archives.gov/infocus/bushrecord/documents/Selected_Speeches_George_W_Bush.pdf.

Chandler, Robert. 2008. *Shadow World: Resurgent Russia, the Global New Left, and Radical Islam*. Washington, DC: Regnery Publishing.

Collins, Michael. 2018. "Indictment: The Russians Also Tried to Help Bernie Sanders, Jill Stein Presidential Campaigns." *USA Today*, February 17.

Conradi, Peter. 2017. *Who Lost Russia? How the World Entered a New Cold War*. Oneworld Publications.

Cull, Nicholas J., Gatov, Vasily, Pomerantsev, Peter, Applebaum, Anne, and Shawcross, Alistair. 2017. "Soviet Subversion, Disinformation and Propaganda: How the West Fought Against It: An Analytic History with Lessons for the Present." London School of Economics, Summary Report (October).

Diamond, Larry, Marc F. Plattner, and Christopher Walker, eds. 2016. *Authoritarianism Goes Global: The Challenge to Democracy*. Baltimore, MD: Johns Hopkins University Press.

Dugin, Alexander. 2014a. *The Fourth Political Theory*. London: Arktos.

Dugin, Alexander. 2014b. *Eurasian Mission*. London: Arktos.

Dunlop, John B. 2004. "Aleksandr Dugin's Foundations of Geopolitics." *Demokratizatsiya* 12, no. 1.

Fish, M. Steven. 2017. "What Is Putinism?" *Journal of Democracy* 28, no. (4): 61–75.

Fisher, Harold Henry. 1955. *The Communist Revolution: An Outline of Strategy and Tactics*. Stanford, CA: Stanford University Press.

Freeden, Michael. 2003. *Ideology: A Very Short Introduction*. Oxford: Oxford University Press.

Galstyan, Areg. 2016. "Third Rome Rising: The Ideologues Are Calling for a New Russian Empire." *The National Interest*, June 27, https://nationalinterest.org/feature/third-rome-rising-the-ideologues-calling-new-russian-empire-16748.

Glazunova, Elena N. 2018. "U.S.-Russian Relations: Dissonance of Ideologies." *Journal of Global Initiatives: Policy, Pedagogy, Perspective* 12, no. (1): 48–67.

Haas, Richard. 2018. "Cold War II." *Council on Foreign Relations Website*, February 23. https://www.cfr.org/article/cold-war-ii.

Hancock, Larry. 2018. *Creating Chaos: Covert Political Warfare from Truman to Putin*. London: OR Books.

Heiser, James D. 2014. "*The American Empire Should Be Destroyed*": *Aleksander Dugin and the Perils of Immanentized Eschatology*. Malone, TX: Repristination Press.

Hiatt, Fred. 2013. "Obama's Broken Commitment to Human Rights in Russia." *Washington Post*, July 15.

IPSOS. 2019. "The Age of Impunity: Global Attitudes to Human Rights." https://www.kcl.ac.uk/policy-institute/assets/age-of-impunity.pdf.

Karyakin, Vladimir. 2013. "Information Wars and Security Threats." *New Eastern Outlook*, October 3.

Kennan, George. 1948. "The Inauguration of Organized Political Warfare." Memo, April 30. https://digitalarchive.wilsoncenter.org/document/114320.pdf?v=941d-c9ee5c6e51333ea9ebbbc9104e8c.

Kintner, William R., and Joseph Z. Kornfelder. 1962. *The New Frontier of War: Political Warfare, Present and Future*. Chicago, IL: Henry Regnery Company.

Kolesnikov, Andrei, and Denis Vertov. 2019. "Pragmatic Paternalism: The Russian Public and the Private Sector." *Carnegie Moscow Center*, January 18. https://carnegie.ru/commentary/78155.

Laqueur, Walter. 2015. *Putinism: Russia and Its Future with the West*. New York: St. Martin's Press.

Laruelle, Marlène. 2012. *Russian Eurasianism: An Ideology of Empire*. Baltimore, MD: Johns Hopkins University Press.

Laruelle, Marlène, ed. 2015. *Eurasianism and the European Far-Right: Reshaping the Europe–Russia Relationship*. Lanham, MD: Lexington Books.

Laruelle, Marlène. 2018. "Is Russia Really 'Fascist'? A Comment on Timothy Snyder." *PONARS Eurasia Policy Memo* 539. http://www.ponarseurasia.org/memo/russia-really-fascist-reply-timothy-snyder.

Lavrov, Sergey. 2019. "Western Liberal Model Is Losing Attractiveness, Lavrov Believes." TASS, April 12, https://tass.com/world/1053401.

Legvold, Robert. 2016. *Return to Cold War*. Cambridge: Polity Press.

Ligerant, Y. 2009. "The Prophet of the New Russian Empire." *Azure* 35, no. 5769.

Lucas, Edward. 2008. *The New Cold War: Putin's Russia and the Threat to the West*. New York: Palgrave.

MacCormac, Sean. 2015. "Aleksandr Dugin: Putin's Rasputin?" *Center for Security Policy*, March 4. https://www.centerforsecuritypolicy.org/2015/03/04/aleksandr-dugin-putins-rasputin/.

Miller, Zeke J. 2014. "Obama on Russia: 'This Is Not Another Cold War.'" Time Magazine, March 26, https://time.com/38988/obama-on-russia-this-is-not-another-cold-war/.

Mills, Curt. 2016. "Panetta: Putin Wants to Restore the Soviet Union." *U.S. News & World Report*, December 1. https://www.usnews.com/news/world/articles/2016-12-01/vladimir-putin-wants-to-restore-the-soviet-union-former-secretary-of-defense-says.

Nance, Malcolm. 2018. *The Plot to Destroy Democracy: How Putin and His Spies Are Undermining America and Dismantling the West*. New York: Hachette Books.

Pejovich, Svetozar. 2018. "From Socialism in the 1900s to Socialism in the 2000s: the Rise of Liberal Socialism." *Post-Communist Economies* 30, no. (1): 117–29.

Perry, B. 2015. "Non-Linear Warfare in Ukraine: The Critical Role of Information Operations and Special Operations." *Small Wars Journal* (August).

Polyakova, Alina. 2014. "Strange Bedfellows: Putin and Europe's Far Right." *World Affairs Journal* 177, no. (3): 36–37.

Pomerantsev, Peter. 2015. *Nothing Is True and Everything Is Possible: The Surreal Heart of the New Russia*. New York: Public Affairs.

Pomerantsev, Peter, and Michael Weiss. 2014. *The Menace of Unreality: How the Kremlin Weaponizes Information, Culture and Money*. New York: Institute of Modern Russia.

Robinson, Linda, Todd C. Helmus, Raphael S. Cohen, Alireza Nader, Andrew Radin, Madeline Magnuson, and Katya Migacheva. 2018. *Modern Political Warfare: Current Practices and Possible Responses*. Santa Monica, CA: RAND.

Schindler, John. 2018. "Russia Has an Ideology—and It's as Entrenched as Communism Was." *The Observer*, March 21.

Schuman, Thomas. 1985. *World Thought Police*. Los Angeles, CA: ALMANAC.

Shachtman, Noah. 2009. "Kremlin Kids: We Launched the Estonian Cyber War." *Wired Magazine*, March 11. https://www.wired.com/2009/03/pro-kremlin-gro/.

Shekhovtsov, Anton. 2016. "How Alexander Dugin's Neo-Eurasianists Have Geared Up for the Russian-Ukrainian War, 2005–2013." *The Interpreter*, January 26. http://euromaidanpress.com/2016/01/26/how-alexander-dugins-neo-eurasianists-geared-up-for-the-russian-ukrainian-war-in-2005-2013/.

Shevtsova, Lilia. 2016. "Forward to the Past in Russia." In *Authoritarianism Goes Global: The Challenge to Democracy*, edited by Larry Diamond, Marc F. Plattner, and Christopher Walker. Baltimore, MD: Johns Hopkins University Press.

Shinar, Chaim. 2015. "The Russian Oligarchs, from Yeltsin to Putin." *European Review* 23, no. (4): 583–96.

Shiner, Chaim. 2018. "Conspiracy Narratives in Russian Politics: From Stalin to Putin." *European Review* 26, no. (4): 648–60.

Shultz, Richard N., and Roy Godson. 1984. *Dezinformatsia: Active Measures in Soviet Strategy*. McLean, VA: Pergamon-Brassey's.

Snegovaya, Maria. 2017. "Is It Time to Drop the F-Bomb on Russia? Why Putin Is *Almost* a Fascist." *World Policy Journal* 34, no. (1): 48–53.

Snyder, Timothy. 2018. *The Road to Unfreedom: Russia, Europe, America*. New York: Tim Duggan Books.

Spiegel Staff. 2014. "The Opinion Makers: How Russia Is Winning the Propaganda War." *Spiegel Online*, May 30. https://www.spiegel.de/international/world/russia-uses-state-television-to-sway-opinion-at-home-and-abroad-a-971971.html.

Steger, Manfred B. 2008. The Rise of the Global Imaginary: Political Ideologies from the French Revolution to the Global War on Terror. Oxford: Oxford University Press.

Steil, Benn. 2018. "Russia's Clash with the West Is About Geography, Not Ideology." *Foreign Policy*, February 12.

Tolstoy, A., and E. McCaffrey. 2015. "Mind Games: Alexander Dugin and Russia's War of Ideas." *World Affairs* 177, no. (6): 25–30.

Tsyngankov, Andrei. 2016. "Crafting the State-Civilization: Vladimir Putin's Turn to Distinct Values." *Problems of Post-Communism* 63, no. (3): 146–58.

Van Herpen, Marcel. 2014. *Putin's Wars: The Rise of Russia's New Imperialism*. Lanham, MD: Rowman & Littlefield.

Van Herpen, Marcel. 2016. *Putin's Propaganda Machine: Soft Power and Russian Foreign Policy*. Lanham, MD: Rowman & Littlefield.

Walker, Shaun. 2014. "Azov Fighters Are Ukraine's Greatest Weapon and May Be Its Greatest Threat." *The Guardian*, September 10.

Washington Times. 2005. "Putin Calls Collapse of Soviet Union 'Catastrophe.'" *Washington Times*, April 26.

Global Security and Intelligence Studies • Volume 4, Number 2 • Fall / Winter 2019

Cyber Force Establishment: Defense Strategy for Protecting Malaysia's Critical National Information Infrastructure Against Cyber Threats

Norazman Mohamad Nor, Azizi Miskon, and Ahmad Mujahid Ahmad Zaidi *National Defence University of Malaysia, Kuala Lumpur, Malaysia*
Zahri Yunos, and Mustaffa Ahmad
Cyber Security Mines, The Mines Resort City, Malaysia

ABSTRACT

Information and Communications Technology (ICT) and the cyberspace have transformed the ways we communicate, power our homes, and obtain government services. Ensuring cyber security requires coordinated efforts from all sectors, both the public and private. Critical National Information Infrastructure (CNII) has become increasingly dependent on ICT, whereby these CNII sectors are interdependent. However, such interdependencies have raised concerns in terms of successful cyber-attacks on certain ICT systems and networks having serious effects on others, resulting in potentially catastrophic disruptions to various services. Therefore, Malaysia requires more expertise to deal with the periodically growing cyber threats, particularly to protect CNII that requires public and private collaboration in cyber crisis management. Malaysia needs to develop a cyber defense strategy that is able to provide adequate protection and response mechanisms at the national level and across CNII sectors. In the event of a cyber crisis, an adequate number of professional and trusted experts are necessary to defend the country. In order to manage this matter, a conceptual framework for the establishment of a cyber force team is proposed. The cyber force team will still work with their present organizations and will only be deployed when there are imminent cyber-attacks or cyber crises at the national level. The team will be equipped with technical skills and knowledge for protecting the country's cyber security. Moreover, the team will be comprised of selected professionals and graduates with an ICT background. A central agency will command the cyber force team to ensure readiness and to faciliate synergies across the public and private sectors that manage the CNII sectors.

Keywords: cyber security, cyber threat, cyber force, critical national information infrastructure (CNII)

 doi: 10.18278/gsis.4.2.4

Establecimiento de la Fuerza Cibernética: Estrategia de defensa para proteger la infraestructura de información nacional crítica de Malasia contra las amenazas cibernéticas

Norazman Mohamad Nor, Azizi Miskon y Ahmad Mujahid Ahmad Zaidi *National Defence University of Malaysia, Kuala Lumpur, Malasia* Zahri Yunos y Mustaffa Ahmad *Cyber Security Mines, The Mines Resort City, Malasia*

Resumen

La tecnología de la información y las comunicaciones (TIC) y el ciberespacio han transformado las formas en que nos comunicamos, alimentamos nuestros hogares y obtenemos servicios gubernamentales. Garantizar la seguridad cibernética requiere esfuerzos coordinados de todos los sectores, tanto públicos como privados. La Infraestructura Nacional de Información Crítica (CNII) se ha vuelto cada vez más dependiente de las TIC, por lo que estos sectores de la CNII son interdependientes. Sin embargo, tales interdependencias han suscitado inquietudes en términos de ataques cibernéticos exitosos en ciertos sistemas y redes de TIC que tienen graves efectos en otros, lo que resulta en interrupciones potencialmente catastróficas para varios servicios. Por lo tanto, Malasia requiere más experiencia para hacer frente a las amenazas cibernéticas que crecen periódicamente, particularmente para proteger la CNII que requiere la colaboración pública y privada en la gestión de crisis cibernéticas. Malasia necesita desarrollar una estrategia de defensa cibernética que sea capaz de proporcionar mecanismos adecuados de protección y respuesta a nivel nacional y en todos los sectores de la CNII. En caso de una crisis cibernética, se necesita un número adecuado de expertos profesionales y de confianza para defender el país. Para gestionar este asunto, se propone un marco conceptual para el establecimiento de un equipo de fuerza cibernética. El equipo de la fuerza cibernética seguirá trabajando con sus organizaciones actuales y solo se desplegará cuando haya ciberataques o crisis cibernéticas inminentes a nivel nacional. El equipo estará equipado con habilidades técnicas y conocimientos para proteger la seguridad cibernética del país. Además, el equipo estará compuesto por profesionales seleccionados y graduados con experiencia en TIC.

Una agencia central mandará al equipo de la fuerza cibernética para garantizar la preparación y facilitar las sinergias entre los sectores público y privado que gestionan los sectores de la CNII.

Palabras clave: seguridad cibernética, amenaza cibernética, fuerza cibernética, infraestructura de información nacional crítica (CNII)

网络力量的建立：保护马来西亚关键国家信息基础设施抵御网络威胁的防御战略

Norazman Mohamad Nor, Azizi Miskon, and Ahmad Mujahid Ahmad Zaidi *National Defence University of Malaysia, Kuala Lumpur, Malaysia* Zahri Yunos, and Mustaffa Ahmad
Cyber Security Mines, The Mines Resort City, Malaysia

摘要

信息通信技术（ICT）和网络空间已改变了人们沟通、为家庭供电、获取政府服务的方式。确保网络安全，要求所有部门协同努力，包括公共部门和私人部门。关键国家信息基础设施（CNII）已越来越依赖ICT，由此CNII各部门相互依赖彼此。然而，这种相互依赖性引发了相关忧虑，即网络攻击成功进入部分ICT系统和网络，对其他网络造成严重影响，并导致对不同服务产生潜在的破坏性干扰。因此，马来西亚要求更多专业技术应对周期性增长的网络威胁，尤其是保护CNII，后者要求公共和私人部门在网络危机管理方面进行协作。

马来西亚需要研发出一种网络防御战略，这种战略能够在全国层面和CNII各部门之间提供具备充分保护和响应的机制。当处于网络危机时，需要足够数量的专家和受信任专业人士共同保卫国家。为管理这一事务，本文就建立网络力量小组提出了一项概念框架。网络力量小组将依旧为现有组织效力，且仅当全国范围内发生紧急网络攻击或危机时才会被调用。该小组将配备专业技术和知识，以保护国家网络安全。此外，该小组成员都是经过挑选的、具备ICT专业背景的专家和研究生。中央机构将指挥该网络力量小组确保做足准备，并促进管理CNII部门的各个公共部门和私人部门发挥协同作用。

关键词：网络安全，网络威胁，网络力量，关键国家信息基础设施（CNII）

Introduction

Cyber seems to dominate and transform our way of life. The ways in which we communicate, interact, and do business have dramatically changed. The impact of this change can be viewed as positive or negative. As information and communication technology has developed and evolved, cyberspace, cyber-crimes, and cyber security must be redefined. Cyberspace refers to the borderless space known as the Internet (Bandara, Ioras, and Maher, 2014). It refers to the interdependent network of information technology components that become the backbone for many of our communications technologies today. Conversely, the negative impacts include cybercrimes which are escalating at an alarming rate in the cyber world. Since computer crime may involve all sorts of crime, a definition must place emphasis on potential uses of computer technology for unethical and illegal purposes. To counter cybercrime, there comes the need for cyber security.

The International Telecommunications Union (ITU) (2008) defines cyber security as the collection of tools, policies, guidelines, security concepts, security safeguards, risk management, best practices, actions, training, assurance, and technologies to manage the cybercrimes. Cyber security will protect the organization and a user's assets in the cyber environment including any connected computing devices, infrastructure, services, and any stored or transmitted information. Against the relevant security risks in the cyber environment, cyber security strives to guarantee the attainment and maintenance of the security properties and assets of an organization. Itari, Anthony, and Mercy (2017) state that cyber security acts as a body of rules established to protect cyberspace. Unfortunately, as we become more dependent on the Internet and cyberspace, we undoubtedly face new risks.

From a larger perspective, cybercrime refers to the series of crime attacking both cyberspace and cyber security (Mosuro, 2017). Sophisticated attacks in cyber space, hacking activities, cybercrimes, among others, present risks to our national security and economy. Malaysia's national security and economic vitality depend on a vast array of interdependent and critical networks in cyberspace. Cyberspace has transformed the ways we communicate, run our economy, run education systems, and obtain government services. Cyber security is the body of technology, processes, and practices designed to protect networks, data, and related devices from attacks, damage, or unauthorized access. In the online world of cyber context, the term *security* simply implies cyber security.

Malaysia needs more expertise to deal with the growing cyber threats, particularly to protect their Critical National Information Infrastructure (CNII) that encompasses public and private collaboration in cyber crisis management. A cyber defense strategy that is able to provide adequate protection and response mechanisms need to be developed at the national level and across CNII agencies or organizations. In the event of a cyber crisis, it will be imperative that a significant number of experts who are professional, loyal, and trusted will be ready to defend our cyber space. Toward that goal, a team of volunteer cyber experts should be organized to serve as a "reserve" force who would be swiftly deployable in the event of an imminent cyber attack or cyber crisis at a national level.

Background of Cyber Force Establishment

Every day, millions of people in the UK depend on cyberspace to perform various activities involving online banking, transactions, and conducting business (UK Office of Cyber Security [OCS], 2009). However, when carried out, each of these activities can be vulnerable to cybercrime. This vulnerability led the UK's government to take preventative measures against cybercrime in March 2014 by establishing a group known as UK's National Computer Emergency Response Team (UK-CERT) in response to the National Security Secretariat (NSS) (HM Government, 2015). The NSS is responsible for coordination on security and intelligence issues of strategic importance across government, including cyber security. The UK-CERT works closely with industry, government, civil liberty groups, and academia to enhance cyber resilience in the UK. The UK's government further developed the UK Cyber Security Strategy (2011–2016) (UK Cabinet Office, 2011) and aims to:

- Make the country one of the most secure places in the world to conduct business online.

- Make the country more resilient to cyber-attacks and to protect interest in cyberspace.

- Help shape an open, stable, and vibrant cyberspace.

- Obtain and grow the crosscutting knowledge, skills, and capability it needs.

In addition to UK-CERT, there are organizations that are also operating in the UK to protect against cyber threats, including the main office of the UK's government, the Serious Organized Crime Agency (SOCA), a cyber security operation center, and the UK's Police to combat criminal activities in cyberspace (UK Office of Cyber Security [OCS], 2009). The NSS engages the people of the UK to help contribute and work together to provide more secure services, operate information systems safely, and protect an individual's privacy. Furthermore, the public has their own responsibility to make sure to protect themselves, and others in society. To help draw interest, a competition called the Cyber Security Challenge, helps identify talented individuals in cyber security (Cyber Security Challenge UK, 2016). The 2016 winner of the Cyber Security Challenge was a postman who currently works as an Information Security Professional for the Royal Mail. In addition to NSS, the UK also has a military cyber defense force led by the UK Ministry of Defence (MOD) that is involved with the military use of cyber space, including defense policy and doctrine (Ministry of Defence [UK], 2014).

In the U.S., there are thousands of cyber-attacks reported everyday that require the government's immediate attention (Weimann, 2004). The U.S. intelligence community reports that terrorism is the number one threat in the United States and has become a contentious issue. Since the early twenty-first century, cyber-attacks occur alongside illegal and malevolent political and financially motivated cyber activity (Sheldon 2012). Unfortunately, the increasing numbers of cyber-attacks used as a political weapon reflects a dangerous, and growing trend in international relations. A disruptive or destructive cyber-attack could be a significant risk to national and global security.

The United States Department of Defence (DoD) is responsible for defending the homeland and related interests from any attack that occurs in cyberspace (The Department of Defense, 2015). Thus, they develop new strategies and prioritized goals and objectives for future DoD's cyber activities and missions. Schmidt (2015) found that effective cyber security and cyber operations focus on building the capabilities to defend the DoD's network, systems, and information; defending the nation against cyber-attacks of significant consequence; and support operational and contingency plans.

The Department of Defence (2015) has carried out a range of activities outside of cyberspace toward improving collective cyber security missions and protecting U.S. interest. The activities include the sharing of information and inter-agency coordination, building bridges to the private sector, and building alliances, coalitions, and partnerships abroad. The DoD's primary mission in cyberspace is to defend and quickly recover its own networks, systems, and information against attacks. The DoD securely operates the Department of Defence Information Network (DoDIN) on an ongoing basis to respond quickly to close vulnerabilities, and secure networks. For the second mission, DoDs seek to synchronize capabilities with other government agencies to develop a range of options and methods for disrupting cyber-attacks of significance consequence. As the U.S. government has limited and specific roles in defending the nation, private sectors and companies are the first line of defense. Their coverage and operations extends to nearly all of the cyberspace networks and infrastructure. For the third mission, the DoD must be able to provide integrated cyber capabilities to support military operations and contingency plans by conducting a United States Cyber Command (USCYBER-COM) to deter or defeat strategic threats in other domains.

Furthermore, a new cyber mission force (CMF) is proposed to appoint multiple leaders and communities to carry out the missions together across DoDs and the U.S. government with commitment and coordination. Personnel involved will have an active role to build and operate DoD networks and information technology systems (The Department of Defence 2015). Thus, all activities have to synchronize between U.S. Cyber Command and other DoD organizations—mainly in combatant commands, to respond to emerging challenges and opportunities. It is highly recommended that the Military Department's Computer Emergency Response Teams (CERTs), DHS, and USCYBERCOM be partnered with the installation owners and operators to build adaptive defenses and continuity plans for mission-critical systems and the civil systems that support them (Theohary and Harrington 2015).

The highlight of this section is the formation of CMF which was built in 2012 by DoDs. It has a unique role within the Department to carry out DoD cyber missions. The CMF is comprised of nearly 6,200 military, civilian, and contractor support personnel from across the military departments and defense components. CMF has been integrated into the larger multi-mission U.S. military force to achieve synergy across the domains. During the process of implementing the strategy, development of CMF has continued to satisfy the necessary command, control, and the enabling of organizations required for effective cyber operations. Concomitantly, DoDs also focus on ensuring that its forces are trained and ready to operate using the capabilities and architectures they need to conduct cyber operations. Further, DODs continue to build policy and legal frameworks to govern CMF employment, and integrate the CMF into DoDs overall planning and force development (The Department of Defense 2015).

The Japanese Cyber Security Strategy was launched in 2013 and the basic mission of NISC (National Information Security Council) is documented in "Information Security 2012." Their main intentions, among many, are to a) strengthen measures to prevent or counter sophisticated threats to companies and organizations handling important national information on security; b) to maintain a safe and secure user environment; and c) to reinforce international alliances. The NISC is the control tower of cyber security which coordinates government efforts and four key agencies: 1) the National Police Agency (NPA); 2) the Ministry of Internal Affairs and Communications (MITI); 3) the Ministry of Economy, Trade, and Industry (METI); and 4) the Ministry of Defence (MOD). NISC also works closely with businesses and individuals. Agencies that are in charge of critical infrastructures, such as the Financial Services Agency, MIC, METI, Ministry of Health, Labour and Welfare, and Ministry of Land, Infrastructure, Transport, and Tourism are required to cooperate with NISC (Nitta 2014).

Among the countries of Southeast Asia, Singapore is a country that succeeded in developing a group and system for the prevention of cybercrimes. The group is known as Sing-CERT and was established in 1997 as a national computer emergency response team and a program within the Infocomm Development Authority of Singapore (IDA). It is established in collaboration with the National University of Singapore (NUS). The functions of this team are more to facilitate the detection, resolution, and prevention of security-related incidents on the Internet.

Additionally, Sing-CERT can also identify hacking trends and activities that can harm any company or agency or that can harm Singapore itself. IDA plays a very important role in planning the formulation of the national Infocomm master plan and policies. They also regulate the telecoms industry and enhance the economic sectors (Ministry of Communication and Information [SG], 2017).

Considering these factors, cIDA intentionally built the plan to require modifications every 3–5 years to ensure that Singapore continues to be a secure and trusted country. The first master plan of IDA was known as Infocomm Security Master plan (ISMP 2005–2007), which included a secured Infocomm environment for government, businesses, and individuals and could defend Singapore's critical infrastructure from cyber-attacks (Cyber Security Agency of Singapore, 2016). Later, they built ISMP 2 (2008–2012) to expand the coverage to include critical Infocomm infrastructure to make Singapore a more secure and trusted hub. However, in April 2015, Sing-CERT moved from IDA to the Cyber Security Agency (CSA). CSA is a computer security agency supporting the Singapore government working under the Prime Minister's Office. The CSA is responsible for ensuring Singapore's cyber security, focusing on strategy and policy development, security operations, industrial development, and working together with private sectors to protect 10 critical sectors including power, transport, telecommunications, and banking from increasing cyber threats (Vu 2016). Even though Sing-CERT moved to CSA, the master plan has continued operations for the last 5 years under the name National Cyber Security Master plan 2018 (NCSM, 2018) to enhance the cyber security for the people of Singapore with six strategies: (1) securing the people, private, and public sectors, (2) developing national capabilities, (3) cultivating technology and R&D, and (4) securing the national infrastructure. In addition to Government and Critical Infocomm Infrastructure (CII), NCSM 2018 are focusing more on conveying security awareness messages via online videos, national television programs, and other communication channels to reach out to businesses and individuals. Additionally, NCSM 2018 also plans to develop (5) more cyber security experts through research and development (R&D), and (6) training and testing programs (Infocomm Development Authority of Singapore [IDA], 2017). To make Singapore more secure, the CSA signed a number of bilateral Memorandum of Understanding (MoUs) with France, UK, and India and they will support the CERT cooperation through the annual ASEAN CERT Incident Drill exercises (Cyber Security Agency of Singapore, 2015). Besides Sing-CERT, Singapore already has a Cyber Defence Operation Hub (CDOH), which includes the Singapore Armed Forces (Lim, 2017). Generally, CDOH is a defense focused more on supporting countries outside of Singapore, while Sing-CERT is a defense force that focuses more within Singapore's boundaries as well as providing the public with a support website in the event of a cyber incident, such as unauthorized attempts, hackings, viruses, and attacks on IT network.

Cyber Security Exercises in Developed Countries

In 2015, the U.S. government launched a program to target security threats that could potentially wipe out all of America's power (Defense Advanced Research Projects Agency [DARPA], 2015). Similarly, on April 2016, the EU also prepared what was known as the "Dark Scenario" which simulated cyber-attacks that could devastate power and communication networks. More than 700 security experts from 30 countries were involved (including IT security companies, bank, energy companies, and CSA) (European Union Agency for network and Information Security [ENISA], 2016). Scenarios featured power cuts, attacks on drones, mobile malware, and ransomware. Experts gave warnings that these cyber-attacks could cause a global catastrophe (including hacks on satellites, spacecrafts, and nuclear power plants).

Meanwhile, in Singapore, CSA has conducted an exercise known as "Exercise Cyber Star" (Ministry of Communication and Information [SG], 2016). This exercise comprises a series of scenario planning sessions, workshops, and table-top discussions focusing on cyber incident management processes. In May 2016, the government of Singapore called for a project worth S$2.82 billion of Information and Communication Technology (ICT) in security tenders across the 2016 fiscal year (FY), according to the IDA of Singapore (Info-Communications Media Development Authority [SG], 2016). Later, within the Singapore 2016 budget narrative, it was stated that the government set aside S$120 million to support training and re-training of Infocomm professionals; focusing on high-demand areas such as software development, data analytics, cyber security, network, and infrastructure for their cyberspace defense (UK-Asean Business Council, 2016). A summary of the Cyber Security Strategy details from other countries such as the UK, the U.S., Singapore, and Japan is shown in Table 1.

Proposed Establishment Models

Cyber Force: A Way Forward for Malaysia

Cyber Force is a force within the National Security Council (NSC) to protect Malaysia's CNII that encompasses public and private collaboration in cyber crisis management. The Forces should be equipped with loyalty, skills, and knowledge in protecting and enhancing the resilience of cyber security for the country. The members of Cyber Force will be selective from Reserve Officer Training Unit (ROTU), professionals, and graduates with Information Technology background. They must undergo patriotism trainings conducted by National Civics Bureau or Malaysia Armed Forces in order to develop trusts and cooperation

Table 1: Summary of Cyber Security Strategy from Other Countries; UK, U.S., Singapore, and Japan

	UK	United States	Japan	Singapore
Policy on Cyber Security	National Cyber Security Strategy (NCSS) (2011–2015)	Department of Defense (DoD) Strategy 2015	Cyber security Strategy (2013)	National Cyber Security Master plan 2018
Engagement of Cyber security professionals and civilian experts with military	• Get Safe Online—public and private sector joint campaign; to raise awareness to public and business • Cyber security challenge competition • Working with people of UK, government, industry, and academia to make UK's cyberspace more secure	Cyber Mission Force (CMF): • Consist nearly 6,200 military, civilian, contractor support • Selected graduated students from the U.S. government's Cyber Corps program	Coordinating government efforts and four key agencies: NPA, MITI, METI, MOD, and work closely with businesses and individuals	• Cyber Security Associates and Technologists Programme (CSAT) • A dual track program: The Associates track Fresh ICT and engineering professionals • The Technologist track Experienced ICT pro and Network Engineers • National Cybersecurity Postgraduate Scholarship
	National Security Secretariat (NCS)	Defense Department	National Information Security Council (NISC)	Cyber Security Agency of Singapore (CSA)

and the professional training by Cyber Security Malaysia (CSM) in order to enhance skills and knowledge for effective operations. NSC will command the forces to develop synergies across domains, ensuring the readiness within the force and coordinating the military and civilian workforce and infrastructure to accomplish the missions.

Initial Concept Proposal for the Establishment of Cyber Force

Figure 1(a) and (b) shows the initial concept to establish the Cyber Force. The first layer of protection will be represented by the law on cyber threats which falls under the jurisdiction of the Malaysian Communications and Multimedia Commission (SKMM). In the second layer of protection, Cyber-Safe will continue to monitor and raise awareness regarding the threats that occur in the cyberspace. When the threat of a cyber-attacks is imminent, at this point, it will be handled by a team from Malaysia Cyber Emergency Response Team (My-CERT) and Malaysia Cyber Security (MyCSC). In this regard, Malaysia needs more expertise to deal with the growing cyber threats. However, to hire a large number of full-time experts will be too costly. In order to overcome this issue, the establishment of a Cyber Force is proposed. Cyber security professionals need to openly and continually communicate to each other to combat cyber threats. During peace-time, this special team would work with their present employers and will only be called upon whenever there are cyber threats or attacks. In order to be effective, cyber security professionals need to be curious, persistent, resilient, ambiguous, and decisive. To ensure the loyalty to the country, experts shall be drawn mainly among qualified alumni of the National Defence University of Malaysia (NDUM) or Royal Military Collage (RMC) where expertise will be evaluated by CSM.

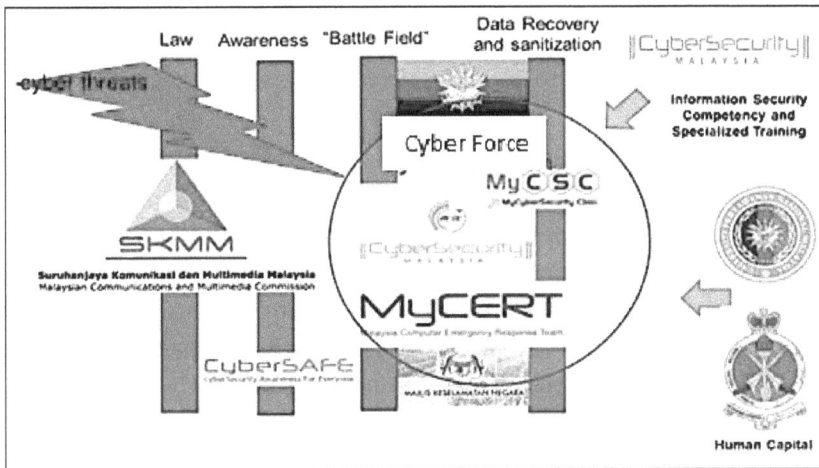

Figure 1(a) The Initial Concept Proposal for the Establishment of Cyber Force

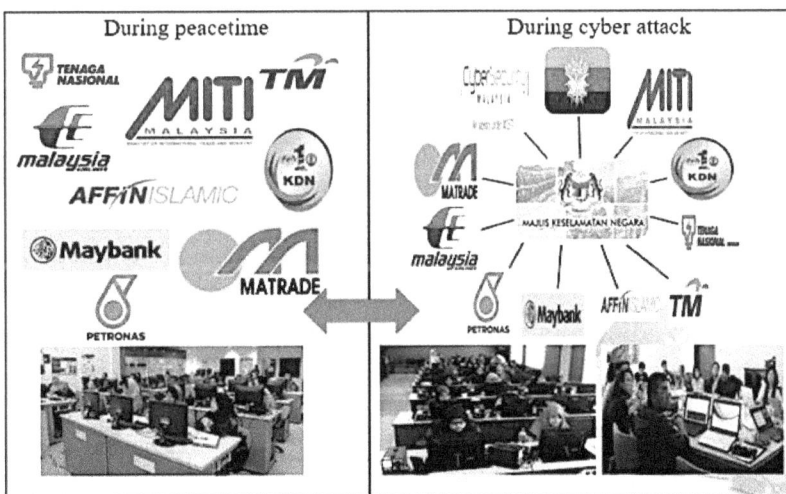

Figure 1(b): The Initial Concept Proposal for the Establishment of Cyber Force

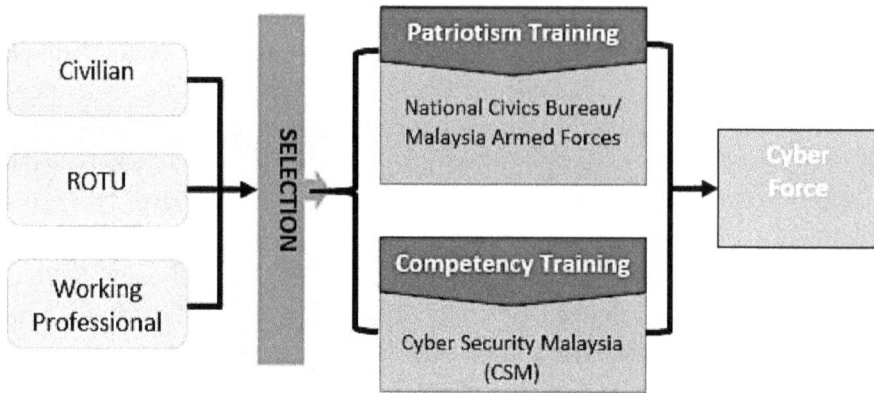

Figure 2: Training Process of Cyber Force

Process of Producing the Cyber Force

As we know, the development of the Cyber Force, who would be individuals chosen among civilian, ROTU, and professional, must be trained before being selected as a member of the Cyber Force. Figure 2 shows the training process of Cyber Force. The structured training will ensure the cyber force members will not only be technically competent but they will be patriotic and trustable. As stated earlier, training in the National Civic Bureau and/or the Malaysian Armed Forces will focus on developing patriotism among the cyber force candidates. Further, the CSM will provide competency training, which will cover various skills to counter threats in cyber space.

Conclusion

The establishment of the Cyber Force concept presents a plan to enhance the readiness level of national cyber security in Malaysia. Success requires close collaboration across relevant agencies of the Malaysian Government, private sector, and with other allies and partners. The plan will start with the establishment of the skeleton organizational structure followed by the recruitment, training, and certification process. Once the force is in place, it would be tested at various levels for readiness and effectiveness. The establishment must be dynamic to address current technologies and current threats in the cyber space.

References

Bandara, I., F. Ioras, and K. Maher. 2014. "Cyber Security Concerns in E-learning Education." Proceedings of ICERI2014 Conference, 728–34.

Cyber Security Agency of Singapore [CSA]. 2015. "CSA Signs MOU with India to Strengthen Cooperation on Cyber Security." https://www.csa.gov.sg/news/press-releases/csa-signs-mou-with-india-to-strengthen-cooperation-on-cyber-security.

Cyber Security Agency of Singapore [CSA]. 2016. *Singapore's Cybersecurity Strategy*. Singapore: Prime Minister's Office (PMO).

Cyber Security Challenge UK. 2016. "Cyber Security Challenge UK Tests Cyber Skills in Virtual Reality in Latest Competition." [Online]. http://www.informationsecuritybuzz.com/articles/cyber-security-challenge-uk-tests-cyber-skills-virtual-reality-latest-competition/.

Defense Advanced Research Projects Agency [DARPA]. 2015. *Breakthrough Technologies for National Security*. Arlington, VA: DARPA.

European Union Agency for Network and Information Security [ENISA]. 2016. "Cyber Europe 2016: The Pan-European Exercise to Protect EU Infrastructures Against Coordinated Cyber-Attack. https://www.enisa.europa.eu/news/enisa-news/cyber-europe-2016.

HM Government. 2015. "National Security Strategy and Strategic Defence and Security Review 2015: A Secure and Prosperous United Kingdom." Presented to Parliament by the Prime Minister by Command of Her Majesty, London, United Kingdom: Williams Lea Group, November.

Infocomm Development Authority of Singapore [IDA]. 2017. *National Cyber Security Masterplan 2018*. Singapore: Government Technology Agency.

Info-communications Media Development Authority [SG]. 2016. "Government and Industry to Build a Smart Nation Together." https://www.imda.gov.sg/about/newsroom/archived/ida/media-releases/2016/government-and-industry-to-build-a-smart-nation-together.

International Telecommunications Union (ITU). 2008. ITU-TX.1205 "Data Networks, Open Sytem Communications and Security: Telecommunication Security: Overview of Cybersecurity." [online]. https://www.itu.int/en/ITU-T/studygroups/com17/Pages/cybersecurity.aspx.

Itari, D., E. O. Anthony, and N. Mercy. 2017. "Cyber Space Technology: Cyber Crime, Cyber Security and Models of Cyber Solution, a Case Study of Nigeria." *International Journal of Computer Science and Mobile Computing* 6, no. 11: 94–113.

Lim, A. 2017. "Singapore Strengthens Cyber Defense with New Organization." http://www.straitstimes.com/singapore/spore-strengthens-cyber-defence-with-new-organisation.

Ministry of Communication and Information [SG]. 2016. "CSA Marks Operational Milestone with Exercise Cyber Star." https://www.csa.gov.sg/news/press-releases/exercise-cyber-star.

Ministry of Communication and Information [SG]. 2017. "About SingERT and How It Can Help You and Your Family." https://www.csa.gov.sg/gosafeonline/go-safe-for-me/homeinternetusers/about-singcert-and-how-can-it-help-you-and-your-company.

Ministry of Defence [UK]. 2014. *United Kingdom Defence Doctrine*. United Kingdom: The Development, Concepts and Doctrine Centre (DCDC).

Mosuro, F. 2017. "Cyber Security—A Matter of Concern for Nigerian Boards." [Online]. https://www.linkedin.com/pulse/cyber-security-matter-concern-nigerian-boards-felicia-mosuro.

Nitta, Y. 2014. "Review of the Japan Cybersecurity Strategy." ISPSW Strategy Series: Focus on Defense and International Security, Issue No. 290.

Schmidt, L. 2015. "Perspective on 2015 DoD Cyber strategy." RAND Office of External Affairs, September.

Sheldon, J. B. 2012. "State of the Art: Attackers and Targets in Cyberspace." *Journal of Military and Strategic Studies* 14, no. 2, 1–19.

The Department of Defense. 2015. *The DoD Cyber Strategy*. Washington, DC: The Department of Defense Cyber Strategy.

Theohary, C. A., and A. I. Harrington. 2015. "Cyber Operations in DOD Policy and Plans: Issue for Congress." Congressional Research Service (CRS) Report (R42848).

UK Cabinet Office. 2011. "The UK Cyber Security Strategy. Protecting and Promoting the UK in a Digital World." (Publication No. 407494/1111). United Kingdom: TSO, November.

UK Office of Cyber Security [OCS]. 2009. "Cyber Security Strategy of the United Kingdom. Safety, Security and Resilience in Cyber Space." Presented to Parliament by the Prime Minister, by Command of Her Majesty, London, United Kingdom: TSO, June.

UK-Asean Business Council. 2016. "Singapore Budget 2016: Renewing Our Economy, Providing Opportunities for All." http://www.ukabc.org.uk/news/renewing-economy-providing-opportunities/.

Vu, C. 2016. "Cyber Security in Singapore." Nanyang Technological Universities, Singapore, Policy Report, December 2016.

Weimann, G. 2004. *Cyberterrorism. How Real Is the Threat*. Washington, DC: United States Institute of Peace.

Acknowledgments

Mr. Neal Kushwaha is the founder and CEO of IMPENDO Inc., a cyber and data center consulting firm in Canada. Annually, he hosts a conference in Ottawa, Canada, called DCAR. During his spare time, he climbs big mountains and will be returning to climb another 8,000 m peak for his spring 2018 expedition in Nepal.

Martti Lehto, Ph.D. (Military Sciences), Colonel G.S. (ret.) is a Professor in Cyber security in the University of Jyväskylä in the Faculty of Information Technology. He has over 30 years' experience as a developer and leader of C4ISR Systems in Finnish Defence Forces. Now, he is a Cyber security and Cyber defense researcher and teacher and the pedagogical director of the Cyber Security M.Sc. program. He is also the Adjunct professor in National Defence University in Air and Cyber Warfare. He has over 100 publications, research reports, and articles on areas of C4ISR systems, digitalization, cyber security and defense, information warfare, and defense policy. Since 2001, he has been the Editor-in-Chief of the Military Magazine.

Wai Sze Leung is an associate professor at the Academy of Computer Science and Software Engineering at the University of Johannesburg. Her current research interests include digital forensics and the application of Artificial Intelligence in enhancing cyber security.

Dr. Andrew Liaropoulos is an Assistant Professor in University of Piraeus, Department of International and European Studies, Greece. He also teaches in the Joint Staff War College, the Joint Military Intelligence College, the National Security College, the Air Staff Command College, and the Naval Staff Command College. Dr. Liaropoulos is also a member of the editorial board of the Journal of Information Warfare.

Candice Louw is a Post-Doctoral Research Fellow appointed at the Department of Business Management (Entrepreneurship) at the University of Johannesburg. Her research interests include smartphone and web application development, technology-driven entrepreneurship, smart infrastructure (cities) management, technological advances in travel and tourism, as well as technology-driven operations optimization strategies and business model development.

Clara Maathuis is a multidisciplinary Ph.D. Researcher in Cyber Operations & Cyber Security at Delft University of Technology, Netherlands Defense Academy, and TNO in the Netherlands. With degrees in Computer Science and Artificial Intelligence, extended her background in military science and military law, and worked as Software Engineer in telecommunications and control systems industry.

Dr. John McAlaney is a Chartered Psychologist, Chartered Scientist, and Principal in Psychology at Bournemouth University in the UK. His research focuses on the social psychological factors of risk behaviors, including cyber security from the perspective of the attackers, the targets, cyber security practitioners, and other stakeholders.

Masombuka Mmalerato is a lead author of the paper and she is currently completing her Master's degree with Stellenbosch University. Second author, Dr. Marthie Grobler is affiliated with CSIRO data 61 in Melbourne, Australia, and Dr. Bruce Watson is with Department of Information Science, Stellenbosch University, and Centre for AI Research, CSIR, South Africa.

Sophia Moganedi is a Cyber Defence Specialist and a Researcher at the Council of Scientific and Industrial Research (CSIR) in South Africa. Her research focus is in the Internet of Things (IoT) from a security and privacy perspective. She has published two peer-reviewed paper in IoT.

Professor Ir Drs. Norazman Bin Mohamad Graduated from University of Texas in 1986 with B.Sc. in Civil Engineering & Mathematics. Then served in the Royal Engineer Regiment of the Malaysian Army. He obtained his M.Sc. from University of Science Malaysia in 1995 and Ph.D. from Cranfield University in 2000. He is currently holding the post of Deputy Vice Chancellor (Research and Innovation) in National Defence University of Malaysia since 2015.

Hussam Mohammed received a bachelor's degree in computer science from University of Anbar, Iraq, in 2009. He was awarded his M.Sc. in computer science from University of Anbar, Iraq, in 2012. He is currently a Ph.D. candidate in the CSCAN at Plymouth University, UK. His research interests include: Big data, security, Digital forensics, and Machine learning.

Global Security and Intelligence Studies • Volume 4, Number 2 • Fall / Winter 2019

Library of Spies: Building an Intelligence Reading List That Meets Your Needs

Erik Kleinsmith,
American Public University System, Charles Town, WV, USA

lifelong learning, intelligence literature, national security, espionage, history of spies and spying, national-level intelligence analysis, biographies, intelligence operations

aprendizaje permanente, literatura de inteligencia, seguridad nacional, espionaje, historia de los espías y del espionaje, análisis de inteligencia a nivel nacional, biografías, operaciones de inteligencia

终生学习，情报文献，国家安全，间谍活动，间谍和间谍活动的历史，国家情报分析，传记，情报活动

Some of the best intelligence professionals are those who would consider themselves life-long learners. These are the people who are persistently driven to professionally improve themselves beyond their formal education and training. A desirable quality in any field, the intelligence candidate or employee who actively seeks to better themselves through new knowledge and experiences is worth more to an organization. Candidates or employees who are life-long learners should be sought out or retained at a higher cost, respectively, as they will guarantee a more rounded, adaptable, and innovative mindset to an intelligence operation.

A significant indicator in finding a life-long learner is in identifying those candidates or employees who are avid readers, especially in intelligence-related topics. Readers of books and other sources related to history, intelligence, national security, espionage, and the more technical aspects of intelligence will bring that wealth of knowledge to their work. In turn, they will be better able to understand and adapt themselves to different problem sets encountered on a regular basis.

For those looking to improve their intelligence-related knowledge base, there are several great reading lists available online including lists published by the CIA, DIA, Marine Corps, and the Small Wars Journal. Each of these provides a good start point for life-long learners to begin expanding their knowledge base. Even with these extensive lists, life-long learners will want to find the right sources

doi: 10.18278/gsis.4.2.5

of literature, prioritizing what to read against what gaps they need to fill in their own knowledge base. They will also need to pick those sources that have the best applicability with their current and expected positions in intelligence.

With this in mind, avid readers will find that sources of intelligence literature can be generally separated into five categories. Each of these categories can be described by their primary focus:

1. **History of Spies and Spying**—Includes covert operations, origins of various intelligence organizations and agencies, historical aspects of spycraft, espionage, and traitors. Most of these more historical subjects are pre-9/11 as they are able to be recounted like a good spy-novel with a beginning, middle, and an end. Historical sources in intelligence are more comprehensive than the most recent stories as they are not currently unfolding or ongoing. For the reader, lessons learned from these sources must either be applied to their current work or kept as a mental note to tap into when a similar situation arises in the future.

2. **National-Level Intelligence**—Includes overviews of the National Intelligence Community and sources focusing specifically on a particular agency or the Intelligence Community in general. Many of these serve as roadmaps for learners to understand the U.S. intelligence landscape at the federal level along with issues and challenges they face. Learners will get their best use of these sources by first starting with those books or sources that provide an overview and then focus in on the particular agencies or strategic level issues that most pertain to the reader's work or interests.

3. **Intelligence Analysis**—Includes sources on analytic techniques, aspects of critical thinking, and case studies using different analytic methodologies. These sources are excellent in helping the intelligence reader expand their "kit bag" of analytic tools that they can use in a given problem they must tackle as part of their job. Most sources on intelligence analysis will list several, walking the reader through each using a lesson-plan format of explaining the technique, walking the reader through the steps, and they showing them how it would work using one or more examples. The key for the intelligence reader with these sources is to understand each of the techniques, but more importantly to be able to apply them to their own tasks and requirements. This could require some significant adaptation in most cases and readers should not be hesitant to break from what is in print in order to meet their own needs.

4. **Biographies**—Includes memoirs and other forms of anecdotal references from national agency directors to operators in the field. These sources can be either historical or current and delve into specific detail about one person's experiences, problems, issues, and means by which they overcame them. Unlike more historical sources covered in the first category, biographical sources tend to be more myopic and from a first-person perspective, not necessarily

in agreement with other perspectives regarding a particular event or issue. Readers need to be very selective and discriminating in selecting sources from this category as several can be seen as rushes to print or "I was there" self-promotions and of questionable accuracy or use.

5. Intelligence Operations—Including intelligence-related doctrine and regulations. These can be military field manuals, circulars, or official government directives and executive orders as well as privately written sources that attempt to tackle intelligence from an organizational and process standpoint. Readers need to understand that many of these sources, especially those that are doctrinal or regulatory, can be somewhat abstract and frankly dry, boring, and somewhat difficult to read from front to back. Best to approach these sources as guides that must be applied to the reader's own problem sets. In full disclosure, I have an upcoming book expected in December of this year that fits in this category.

In addition, life-long learners should also seek out other sources where intelligence may not be the primary subject, but provide a broader topic for the reader to apply what they know about intelligence from their own learning and experience. These could be sources about a particular military campaign, or persistent issues such as gangs, drugs, or cyber threats. Intelligence readers should look to read these sources intermittently with intelligence-specific sources in order to balance intelligence concepts with the skills required to apply and adapt them to their own work.

Network Science in Intelligence: Intelligence Cell

Romeo-Ionuț Mînican
National Defence University of Romania, Bucharest, Romania

ABSTRACT

The world of secret services is governed by unwritten rules in their mission HUMINT alongside SIGINT, OSINT, GEOINT, TECHINT, and cyber intelligence are the base domains for a secret service to obtain information. At the core of intelligence and national security is the human being. Humans evolve throughout life, through education and daily events that occur. Firstly, humans evolve through the education offered by parents, grandparents, etc. Secondly, humans develop skills and competencies in formal education institutions. During this stage, humans are influenced by their own affinities. Thirdly, one of the most important forms of education for humans is that of real-life experiences, interhuman relations, and continual learning. For intelligence officers, these last forms of education significantly contribute to professional career growth. Having training in graph theory, game theory, managing transition, and transactional analysis, for example, are all areas in which intelligence officers will always need continual, lifelong learning.

"Scientia potentia est." Therefore, I argue that in the field of intelligence (also in counterintelligence) and national security, graph theory can be used as a basic hypothesis to identify the intelligence cell, information networks, the base circuit of information, the structure of secret services, the organization and the arrangement of safe houses, and the composition of network of informants, internal or external, on national, neutral, or foreign territory. And some of the most important, and yet neglected, are the weak links and the joker man.

Keywords: cyber intelligence, cybersecurity, network intelligence, intelligence cell, graph theory, game theory

doi: 10.18278/gsis.4.2.6

Ciencia de redes en inteligencia: célula de inteligencia

Romeo-Ionuţ Mînican
National Defence University of Romania, Bucarest, Rumanía

Resumen

El mundo de los servicios secretos se rige por reglas no escritas en su misión. HUMINT junto con SIGINT, OSINT, GEOINT, TECHINT y la inteligencia cibernética son los dominios básicos para que un servicio secreto obtenga información. En el núcleo de la inteligencia y la seguridad nacional está el ser humano. Los humanos evolucionan a lo largo de la vida, a través de la educación y los eventos diarios que ocurren. En primer lugar, los humanos evolucionan a través de la educación ofrecida por los padres, abuelos, etc. En segundo lugar, los humanos desarrollan habilidades y competencias en las instituciones de educación formal. Durante esta etapa, los humanos están influenciados por sus propias afinidades. En tercer lugar, una de las formas más importantes de educación para los humanos es la de las experiencias de la vida real, las relaciones interhumanas y el aprendizaje continuo. Para los oficiales de inteligencia, estas últimas formas de educación contribuyen significativamente al crecimiento profesional. Tener capacitación en teoría de gráficos, teoría de juegos, gestión de la transición y análisis transaccional, por ejemplo, son todas áreas en las que los oficiales de inteligencia siempre necesitarán un aprendizaje continuo y de por vida.

„Scientia potentia est." Por lo tanto, sostengo que en el campo de la inteligencia (también en contrainteligencia) y la seguridad nacional, la teoría de grafos puede usarse como una hipótesis básica para identificar la célula de inteligencia, las redes de información, el circuito base de información, la estructura de los servicios secretos, la organización y la disposición de las casas de seguridad, y la composición de la red de informantes, internos o externos, en territorio nacional, neutral o extranjero. Y algunos de los más importantes, y aún descuidados, son los eslabones débiles y el comodín.

Palabras clave: ciberinteligencia, ciberseguridad, inteligencia de red, celda de inteligencia, teoría de grafos, teoría de juegos

网络情报科学：情报小组

Romeo-Ionuț Mînican
National Defence University of Romania, Bucharest, Romania

摘要

情报组织的世界受其使命中不成文的规则管理，使命除了人力情报外，还有信号情报、开源情报、地理空间情报、技术情报。同时网络情报是情报机关获取信息的基本领域。情报和国家安全的核心则是人。人在一生中通过教育和日常事件不断发展。第一，人通过父母、祖父母等提供的教育得以发展。第二，人在正式教育机构中发展技能和能力。在此期间，人受与其亲近的关系影响。第三，对人类而言最重要的教育形式之一则是现实生活的经历、人与人之间的关系、以及持续学习。对情报官员而言，这些教育的最终形式显著促进了其专业事业的发展。例如图论、博弈论、过渡管理、沟通分析等方面的培训，这些都需要情报官员进行持续不断的终生学习。

"知识就是力量"。因此我主张，在情报（以及反情报）和国家安全领域，图论能被用作为一个基本假设，以识别情报小组、信息网络、信息基本回路、情报机关结构、安全避难所的组织和布置、以及在国家领土、中立领土、或外国领土的内外部告密者网络的构成。并且一些最为重要却被忽视的部分则是薄弱环节和情报小组中的指挥者（joker man）。

关键词：网络情报，网络安全，互联网情报，情报小组，图论，博弈论

Introduction

Graph theory together with game theory can offer policymakers not only a new way of thinking, but also new perspectives on intelligence and national security. When discussing improvements to national global security, fundamental concepts to keep in mind is the differentiation of the "country image," and the "city image," in terms of urban security.

From a micro-level to the macro-level, graph theory can assist in identifying an intelligence cell or more, an intelligence network or more, a network of informants or more, etc. By identifying the intelligence cell in an institution or the intelligence network, it is possible to identify, with a higher accuracy, the people who are in the position of influence and those who lead the institution from the shadow. The construction and network structure is one of the keys to understanding the complex world we live in (Barabási 2002, 12).

Using graph theory in network intelligence analyses offers a key feature represented by the links and connections between nodes. The links between intelligence officers indicate the information flow, the structure of the intelligence cell, network of informants, safe houses, and the meeting places to discuss tête-à-tête.

Regarding the intelligence cell, it is composed of a minimum of three to a maximum of seven people. Depending on the case, the cell can be made up of 12 or 13 people. Taking into account the size of the public or private entity, the intelligence cell or the intelligence network may assume the basic core structure and the architecture represented by Fibonacci's golden spiral.

Therefore, I propose that when conducting intelligence analyses, whether with a public or private entity, company, agency, or a secret service, the intelligence cell be represented as a graph composed of either intelligence officers belonging to the same secret service, or several secret services. Depending on the mission, it can also be set up in a bureau/department/section/division. The structure of an intelligence cell, regardless of size and its hierarchy, are based on the principle of the wolf pack. The alpha male is the leader, or, as the case may be, the alpha female. Using this principle, we can classify the place and the role of intelligence officers to figure out who is the leader and who *really* leads. This example sets out a basic rule in the field of creating and organizing intelligence cells and networks.

As a working hypothesis, the basic structure can be represented by a graph or it can be made up of several graphs, or it can be composed of several graphs until it reaches the real network that leads. The nodes represent either intelligence officers, counterintelligence officers, informants, or there may be cases when nodes represent a network itself. In this case, there is a network into a network, and the link is given by a double node. This hypothesis is a basic framework in identifying intelligence cells or networks. Its representation is based on operational field information to blueprint different networks. In the next stage, the quality of each node is determined: who is the intelligence officer, who are informants, who is counterintelligence; the hierarchy of leadership and the base core, as well as its links with other base core within or outside the secret service. Analogue to the adjacency matrix of the graph, a matrix can be developed whereby those involved and their respective responsibilities are noted, as well as the status of the links between them. This matrix refers to identifying the links between the people involved in the network and possible external links with other networks.

In graph theory, there exists the basic rule that states: "*All nodes with an odd number of links are both a starting point and a stopping point of the route*" (Barabási 2002, 12). The base core is the starting point and the arrival of information within the information circuit. But within the information circuit can occur, intentionally or unintentionally, disturbances or black swan events. The other nodes represent intelligence support nodes for the base core node. In a world where information goes beyond any boundary, it can be appreciated that *"not all information is knowledge—quantity of knowledge does not equal quality [...]"* (Mullins and Christy 2010, 192). Hence, the difficulty in finding and differentiating the real information from propaganda, rumors, gossip, fake news, or public or private opinion.

Similar to the intelligence cell and intelligence networks, graph theory can be used to identify safe houses in the urban area, as well as secret service base offices. By using surveillance, you can find the road or circuit of information from man to man. Due to the formidable technological advance, intelligence officers turn to the old methods of transmitting information. Thereby, the identification of links is essential. Links can be local or foreign informants. Regarding to safe houses in a city, they can be arranged as a graph or more interconnected graphs, depending on the size and strategic importance of the city.

A basic point in identifying a safe house on the city map: the house is located at the intersection of at least two streets. It is also important to know the smugglers' route to get rid of police officers in secret actions. Thus, a node can be a basic safe house where intelligence officers meet with other intelligence officers or hierarchical superiors. This is a basic node of the service in the territory. But a node can be a safe house where intelligence officers meet with agents or informants. In an emergency or unforeseen situation, it can turn into a "fake" safe house, to induce misleading information or to provide a wrong route to enemies. One of the nodes is the core nucleus that coordinates the other nodes (safe houses) in a city or metropolis.

Nowadays, intelligence officers have many ways to communicate and travel (they can have multiple and varied routes) compared to the past century, but the head of the intelligence center has only one point of departure and arrival, and that is the base from where the intelligence cell or network, internal or international, is headed. Any form of organization and leadership is structured and hierarchized in the form of a pyramid. Therefore, the power within a secret service is organized in the form of a pyramid. Similarly, any intelligence cell or network has the power in the form of a pyramid. Using graph theory and game theory in intelligence and national security, the secret services as well as the ultra-secret services can be identified, their intelligence officers, and their location on the territory of a city and country.

When identifying and blueprinting a network and its hierarchy, I argue a basic rule: each network has its own hierarchy of leadership. Therefore, human beings are part of as many hierarchies as the networks they belong to. When did the networks appear? In my opinion, they emerged with the beginning of world trade. Consequently, networks have become active through the exchange of goods, services, information, science, and so on. The activity of all networks is more intense than usual in the transitional years.

Another issue is the concept known as *"six degrees of separation."* It is no longer up to date for the whole world, as all the data of the problem has changed. Starting from the idea of Frigyes Karinthy, from the experiment conducted by professor Stanley Milgram (Barabási 2002, 26–29) and considering the size of the world's population, using the scale of 1: 1,000,000,000, I propose a possible number of eight people to reach any person on the planet.

Therefore, you can reach any person around the globe through at least eight people, two of which are personal acquaintances, one of which is a personal acquaintance, and the other serves as a mentor. Personal acquaintance refers to family and friends, and mentoring type refers to having at least one mentor. Therefore, you need eight people to get to the octagonal personal chart of the person you want to reach. An exception may be the case in which you can reach directly through a direct link. The more links added, the distance to the person we want to reach decreases. Therefore, in order to discover the information on a network, we must add as many links as possible, whether they are temporary, to destructure the network or to fully control it. With regard to cyber security, the power of a virtual space network also consists of its links, i.e. uniform resource locators (URLs). Due to the fact that human social networks intertwine with the web through social networks, if we use the two networks together, the distance between two randomly chosen people will decrease.

In the institutional analysis of intelligence, in addition to the audit itself, it is also necessary to identify which networks are within the institution and what the links are. It is necessary to ascertain which bureau/department/section/division, or intelligence structure acts, following the identification of the informants and the intelligence officers covered. Also, in order to have a complete picture, the social, political, or economic links of people in the institution with those outside the institution must be known.

Therefore, I conclude by emphasizing a basic rule: always personal link will triumph over professional links. There are, of course, exceptions, but rarely encountered. Weak links play a key role in communicating with the outside world of its own personal network of family and friends. An extremely important component of our social network is *the connectors*. They are the thread of society that bring together diverse groups, and people with different origins and training levels (Barabási 2002, 56). There are cases, when one individual has an immense number of links can therefore be a human connector hub.

Within the intelligence cell and in intelligence network, an essential feature is the trust-authority report. The weak links can become strong if they are united. Analogue, the weak networks if they are united, will become strong and pose a threat to the network core, or if the weak links of the networks will be united, it will result in the elimination of some nodes.

Recommendations

The Community Intelligence has to face the major problem of all types of network vulnerability caused by interconnectivity. A first solution would be the development of auxiliary networks (back-up), followed by ensuring the resilience of critical communications infrastructures. With regard to social networks, the weak links of the networks may be their survival. Identifying vulnerable links and nodes of networks must be a constant concern of the secret services. The new major challenge for a secret service is the detection of auxiliary networks (back-up), both for the operative and analysis (intelligence).

Using graph theory and game theory must be a paramount for every secret service in order to constantly adapt and cope with the unpredictable. Although the concept of "six degrees of separation" is out of date, it can be applied regionally.

Another solution for the Intelligence Community is to use not only the social web but also the online social networks and to draw up maps showing the links and the status of friendships, in order to identify the weak links, who is the joker man and who is the main man in each network or intelligence cell.

References

Barabási, Albert-László. 2002. *Linked. The New Science of Networks*. Cambridge, MA: Perseus Publishing.

Baran, Paul. 1964. *On Distributed Communications: I. Introduction to Distributed Communications Networks*. Memorandum RM-3420-PR, August 1964. Santa Monica, CA: The RAND Corporation. Accessed January 19, 2019. https://www.rand.org/pubs/research_memoranda/RM3420.html

Ferguson, Niall. 2018. *The Square and the Tower: Networks, Hierarchies and the Struggle for Global Power*. Bucharest: Polirom Publishing House.

Mullins, Laurie J., and Gill Christy. 2010. *Management & Organizational Behavior*, 9th edition. Harlow, Essex, England: Financial Times Prentice Hall.

Richer, Martha. 1953. *My Espionage Activity*. Russian Book Publishing House.

Global Security and Intelligence Studies • *Volume 4, Number 2* • *Fall / Winter 2019*

Review of *Humanitarian Aid, Genocide, and Mass Killings: Medecine Sans Frontieres, the Rwandan Experience, 1982–97*

Justin West

Harvard University, Cambridge, MA, USA

Jean-Hervé Bradol, Marc Le Pape (2017). *Humanitarian Aid, Genocide, and Mass Killings: Medecins Sans Frontieres, the Rwandan Experience, 1982–97*. Manchester University Press. ISBN: 978-1-7849-9305-4. 160 pages. £80.00

Following periods of mass murder accompanied by myriad humanitarian crises, a question faced by nongovernmental organizations (NGOs) like Medecins Sans Frontieres (MSF) who were operating in Rwanda during the early 1990s is "where were you, and what were you doing?" Authors Dr. Jean-Hervé Bradol and sociologist Marc Le Pape seek to answer the question in two parts. First, how were MSF's activities implemented in an environment of mass murder, military operations, political upheavals, and forced population displacements? Second, in these types of situations, what were the salient or reference points that aid workers called on to guide their actions? This book is not intended to endorse or condemn any particular actions by NGOs, nor to fill informational gaps in archives, but rather to give an account of the ordeals experienced by NGO workers, and how they lived through them. Having served extensively within MSF and performed work in Rwanda in the 80s and 90s, the authors are able to provide context to the troves of documents used to compile this book. The content of messages between the field and home offices, meeting reports, situation and strategy analysis, press releases and reports, and other documents containing data provide the bulk of the text.

To provide historical and political context, the book begins with a look at the years 1980–1994 and describes the birth of MSF aid programs in the region. The relationship of the Kinyarwanda-speaking Hutus and Tutsis to one another, and to the other countries in the region, is described in relation to the call for humanitarian aid that conflicts generated. The reader is able to gain perspective on the political and cultural tensions that culminated in the killing of an estimated 800,000 Tutsis in the span of 100 days, and the deaths of so many Hutus and Tutsis before and after. The resulting mass migrations and health crises were nothing short of disastrous. The core mission of MSF was organizing and providing medical care to displaced people, only in this case the displaced people were being hunted, which led to MSF and other NGO workers witnessing a disproportionate

        doi: 10.18278/gsis.4.2.7

number of atrocities. This trend continued on a lesser scale as the power shifted, the only thing that changed were the names of the victims and the groups responsible. After the Tutsi Rwandan Patriotic Front (RPF) defeated the Hutu government in July 1994, it was the Hutu who were displaced, taking refuge in neighboring nations Zaire, Uganda, Tanzania, and Burundi. The authors continue to present MSF accounts of humanitarian crises and human rights abuses, corresponding to their work in refugee camps and elsewhere through 1997. Also prevalent throughout the book are sets of data related to medical studies, such as relate to morbidity and mortality rates, the nutritional well-being of children, as well as logistical aspects of MSF medical programs. A dozen maps and hundreds of descriptive footnotes allow the reader to reference handwritten notes and datasets, related to everything from outbreaks of disease to firsthand accounts of genocide. Many of the accounts relayed feel close to home, as over 200 MSF employees were murdered because of their ethnicity, which is palpable in the delivery by the authors.

The story is personal, and a bit of expected if not understandable bias may be present, despite the authors' attempts to prevent it. While this may not change the truthfulness of the writers' accounts, nevertheless, it is something the reader should take into consideration. The authors are freer to critique the United Nations High Commissioner for Refugees (UNHCR) activities, for example, and seem less willing to criticize MSF actions. There is also very little chronology to the order of the manuscript, making it very difficult to maintain any sort of timeline to the events described. In one paragraph, you might be expected to follow back and forth between three or more separate years, only to begin in a different decade in the next paragraph. While chronological order is not always a necessity, it seems to be given up with no identifiable attribute gained. In order to enjoy the robustness of this book, the reader must pause often to reassess how an account of a specific event fits into the overall narrative.

All told, this book is a must-read for anyone interested in the field of humanitarian aid or international security. It not only provides a thorough perspective of one the NGOs closest to the events surrounding the Rwandan genocide of 1994, it also explores some previously uncharted territory of the role of an NGO in such a treacherous time and helps the reader to attempt to answer the questions MSF workers have asked themselves in the aftermath of these ghoulish events. How do you comprehend, during an emergency, the political and social dynamics unique to each situation of extreme violence? How do you avoid falling victim to or becoming accomplices of criminal forces? How do you remain effective in such situations? These are some questions that frontline NGOs must ask themselves, and this book may go a long way in helping uncover answers in the future by learning from the past. It would be a good addition to any university curriculum, or useful as part of training for a humanitarian aid organization's leadership.

Global Security and Intelligence Studies • *Volume 4, Number 2* • *Fall / Winter 2019*

Review of *International Organizations and the Law*

Joel Wickwire
American Public University System, Charles Town, WV, USA

Andrea R. Harrington (2018). *Organizations and the Law*. Routledge. ISBN: 978-0815375319. 290 pages. (Kindle) $36.07 (Hardcover) $88.19

Andrea R. Harrington's book, *International Organizations and the Law*, conducts an encyclopedic examination of the nature of international organizations (IOs), providing a comprehensive investigation of historical and contemporary IOs that comprise the world's increasingly controversial security apparatus. One of the most important themes throughout the text is the notion of legitimacy. Harrington argues that for IOs to retain their legitimacy on the world stage, they must exhibit "elements of openness and flexibility, and an understanding of the law and rules that are applicable..." (275). IOs are extremely complicated entities that perform a vast number of services within many different jurisdictions. The ability to adapt and evolve, Harrington argues, is key to their utility and survival.

IOs often promote norms that are eventually adopted in post-conflict areas for missions of reconstruction. However, as is still the case, nations hold their own sovereign interests above those of other nations' interests. This makes the international environment one based on agreements and multilateralism, placing IOs in a difficult situation that can lead to their demise. Harrington highlights IOs in a number of areas including governance, legal, and human rights (United Nations), health (WHO), finance (WTO), and security (NATO). NATO provides a good example of an organization that was endured shifts in the geopolitics through amending its mandate. Harrington cites NATO's recent admittance of Eastern Europe countries as having required "altering the initial vision of NATO in terms of membership and geographic jurisdiction" (33).

In order for IOs like NATO to adapt to evolving threats, it is often required for them to make fundamental amendments to their "foundational texts." It is the area of organizational introspection where Harrington really contributes to one's understanding of how IOs function. For NATO, evolving meant offering membership to a broader range of states, but it also meant a shift in threat assessment, from the narrow focus on Soviet-related threats to looking more generally at criminal activity like trafficking and cybercrime. Indeed, today NATO even supports social programs that promote a spectrum of principles such as rule of law and democracy in nonmember Eastern European countries like Moldova. This collective notion of security is echoed in Harrington's insightful example of The Organization for the Prohibition of Chemical Weapons that was created to protect military personnel from such attacks.

 doi: 10.18278/gsis.4.2.8

Harrington lauds the International Criminal Court (ICC) as being particularly innovative. She explains that the ICC is: the first sustained international court for the prosecution of jus cogens-based crimes, specifically genocide, war crimes, crimes against humanity and the crime of aggression. The ICC is a unique entity with a similarly unique system of governance, one that is persistently challenged by Member States, especially those threatening to leave the ICC based on its prosecutorial decisions. From an apolitical perspective, the ICC represents an innovation and one of the newer forms of international organization. (6)

One reason provided by Harrington for why member states might threaten to leave is due to disagreements of the definition of the ICC's core crimes that fall under its jurisdiction: crimes against humanity, genocide, war crimes, and the crime of aggression. It wasn't until 10 years after the establishment of the Rome Statute that there was an amendment to define the crime of aggression. The alteration to the Rome Statue was called Kampala Amendment, an action that illustrates the ability of the ICC to effect internal change. The ICC is unique in that it prosecutes individuals, it does not allow for immunity, and its mandate for regular dialogue between state/branches/committees. Additionally, within the Rome Statute, there is a provision that has established the Trust Fund for Victims that aid victims of crimes under the Court's jurisdiction.

The UN, ICC, and Interpol all have agreements that facilitate information exchange and in fact, Interpol conducts fact-finding missions for the ICC, as Harrington explains. Interpol is structured and governed like many other IOs, in that each member is given a single vote in decision-making bodies and governing organs, but this right may be suspended in the event the state does not pay its dues. Interpol, as Harrington explains, is unique because it is not a "policing force" so much as it facilitates the "coordination and cooperation between States' policing entities in order to allow for the application of international laws and principles in a more uniform manner" (229).

In this way, Interpol is naturally flexible and able to adapt to any member state through contact with its local Secretariat. Harrington notes Interpol as an example of an IO that was literally formed "in an effort to create a strong network of police entities that can effectively combat crime in multiple and evolving forms" (Ibid). Harrington explains that one-way Interpol cooperates with other IOs like the UN or the ICC is through official agreements. Agreements have established Interpol's international networking system as a primary tool for communicating sanctions and the issuing of notices that effectively act as international arrest warrants. Agreements can cover a wide range of commitments and can specify cooperation on a narrow type of training over a short period of time or agreements can be opened ended and focus on general cooperation.

In her book, Harrington provides an extremely detailed examination of the internal governance of IOs that enables them to function and retain legitimacy within the international environment. The primary focus here was legal and security organizations, but Harrington offers equally impressive insight on organizations in the health and finance sectors. Her exploration of diplomatic rights versus IO rights, what goes into to establishing headquarter agreements, and human rights and environmental organizations is thorough and important for any international scholar or practitioner to understand when researching or working with and/or between these organizations. This book is useful for all who are first learning about IOs or for those who need to refamiliarize themselves with the nature of these organizations that have come to provide a general sense of governance in the international community.

www.ingramcontent.com/pod-product-compliance
Lightning Source LLC
Chambersburg PA
CBHW061618210326
41520CB00041B/7487